AF796413

STEMD^2 Research & Development Group
Center on Disability Studies
College of Education
University of Hawaiʻi at Mānoa

http://stemd2.com/

 First published (2019) by the Center on Disability Studies, University of Hawaiʻi at Mānoa, Honolulu, Hawaiʻi. This document is based upon work supported by the Department of Education, Native Hawaiian Education Act Program under award #S362A180011. Any opinions, findings, and conclusions or recommendations expressed in this material are those of the authors and do not necessarily reflect the views of the United States Department of Education. For further information about this document and Neʻepapa Ka Hana 2.0 project, please contact Dr. Kaveh Abhari at abhari@hawaii.edu.

ISBN: 978-0-9983142-8-0
First release, 2019

Ne'epapa Ka Hana Seventh-Grade Mathematics Resources

Let's Take Care of the Lo'i

Student Activities

Project Director	Kaveh Abhari
Content Developers	Robert G. Young Justin S. Toyofuku
Creative Designer	MyLan Tran
Publication Designer	Robert G. Young

Acknowledgments

We would like to thank Kelli Ching, Crystal Yoo, Katy Parsons, and Robyn Rice for advising on middle school mathematics. Thank you Moa Viebke, Nohea Behler, and Robyn Rice for significant help reviewing and editing. Mahalo nui to Moa Viebke and Nohea Behler for major contributions in writing the introductions.

Contents

Introduction to *Let's Take Care of the Lo'i*

Let's Take Care of the Lo'i has been created by Ne'epapa Ka Hana (NKH) for 7th grade students of Hawai'i. The mathematical activities featured in *Let's Take Care of the Lo'i* have been constructed to enrich students mathematical abilities through culturally responsive material and to increase interest and participation in the mathematical classroom.

Through the collaboration between the Ne'epapa Ka Hana (NKH) 2.0 and STEMD2 the mathematical curriculum of *Let's Take Care of the Lo'i* has been made available for online classroom integration. The math activities are offered on a social platform which allows for online collaboration between students and teachers. The platform is free and accessible to users with internet access at www.community.stemd2.com

Ne'epapa Ka Hana (NKH) develop programs and materials that are culturally responsive to Hawai'i's unique diversity – using research-based practices to support ongoing STEM education efforts across the state. *Let's Take Care of the Lo'i* is the 7th addition to NKH's middle-school book series.

The *Let's Take Care of the Lo'i* book is composed by mathematical learning activities that serve as a supplementary mathematics curriculum to enhance inclusive instruction based on problem-based learning strategies and connectivism principles. The math activities incorporate real-world challenges that reflect Hawai'i's unique culture, society, and geography. The goal of these activities is for both kumu and haumāna to collaborate while thinking critically and creatively – thereby, deepening students' understanding, application, and appreciation for mathematical thinking in the context of Hawaiian and island culture.

Let's Take Care of the Lo'i, highlights the Common Core State Standards through the theme of food, farming, and fishing in Hawai'i. The curriculum focuses on supporting food systems in Hawai'i and providing a background of the food that sustained the Hawaiian population through mathematical problems.

Common Core State Standards

Common Core State Standard	*Let's Take Care of the Lo'i* Unit
Ratios & Proportional Relationships	
7.RP.1 Compute unit rates associated with ratios of fractions, including ratios of lengths, areas and other quantities measured in like or different units. For example, If a person walks 1/2 mile in each 1/4 hour, compute the unit rate as the complex fraction (1/2)/(1/4) miles per hour, equivalently 2 miles per hour.	2
7.RP.2 Recognize and represent proportional relationships between quantities.	2
7.RP.2a Decide whether two quantities are in a proportional relationship, e.g., by testing for equivalent ratios in a table or graphing on a coordinate plane and observing whether the graph is a straight line through the origin.	2
7.RP.2b Identify the constant of proportionality (unit rate) in tables, graphs, equations, diagrams, and verbal descriptions of proportional relationships.	2
7.RP.2c Represent proportional relationships by equations. For example, if total cost t is proportional to the number n of items purchased at a constant price p, the relationship between the total cost and the number of items can be expressed as t = pn.	5
7.RP.2d Explain what a point (x, y) on the graph of a proportional relationship means in terms of the situation, with special attention to the points (0, 0) and (1, r) where r is the unit rate.	2
7.RP.3 Use proportional relationships to solve multistep ratio and percent problems. Examples: simple interest, tax, markups and markdowns, gratuities and commissions, fees, percent increase and decrease, percent error.	2

Continued on next page

Common Core State Standard	***Let's Take Care of the Lo'i* Unit**
The Number System	
7.NS.1 Apply and extend previous understandings of addition and subtraction to add and subtract rational numbers; represent addition and subtraction on a horizontal or vertical number line diagram.	1
7.NS.1a Describe situations in which opposite quantities combine to make 0. For example, a hydrogen atom has 0 charge because its two constituents are oppositely charged.	1
7.NS.1b Understand p + q as the number located a distance |q| from p, in the positive or negative direction depending on whether q is positive or negative. Show that a number and its opposite have a sum of 0 (are additive inverses). Interpret sums of rational numbers by describing real-world contexts.	1
7.NS.1c Understand subtraction of rational numbers as adding the additive inverse, p − q = p + (−q). Show that the distance between two rational numbers on the number line is the absolute value of their difference, and apply this principle in real-world contexts.	1
7.NS.1d Apply properties of operations as strategies to add and subtract rational numbers.	1
7.NS.2 Apply and extend previous understandings of multiplication and division and of fractions to multiply and divide rational numbers.	1
7.NS.2a Understand that multiplication is extended from fractions to rational numbers by requiring that operations continue to satisfy the properties of operations, particularly the distributive property, leading to products such as (−1)(−1) = 1 and the rules for multiplying signed numbers. Interpret products of rational numbers by describing real-world contexts.	1
7.NS.2b Understand that integers can be divided, provided that the divisor is not zero, and every quotient of integers (with non-zero divisor) is a rational number. If p and q are integers, then −(p/q) = (−p)/q = p/(−q). Interpret quotients of rational numbers by describing real-world contexts.	1
7.NS.2c Apply properties of operations as strategies to multiply and divide rational numbers.	1
7.NS.2d Convert a rational number to a decimal using long division; know that the decimal form of a rational number terminates in 0s or eventually repeats	1

Continued on next page

Common Core State Standard	***Let's Take Care of the Lo'i* Unit**
7.NS.3 Solve real-world and mathematical problems involving the four operations with rational numbers. (Computations with rational numbers extend the rules for manipulating fractions to complex fractions.)	1
Expressions & Equations	
7.EE.1 Apply properties of operations as strategies to add, subtract, factor, and expand linear expressions with rational coefficients.	3
7.EE.2 Understand that rewriting an expression in different forms in a problem context can shed light on the problem and how the quantities in it are related. For example, a + 0.05a = 1.05a means that "increase by 5%" is the same as "multiply by 1.05."	2
7.EE.3 Solve multi-step real-life and mathematical problems posed with positive and negative rational numbers in any form (whole numbers, fractions, and decimals), using tools strategically. Apply properties of operations as strategies to calculate with numbers in any form; convert between forms as appropriate; and assess the reasonableness of answers using mental computation and estimation strategies. For example: If a woman making $25 an hour gets a 10% raise, she will make an additional 1/10 of her salary an hour, or $2.50, for a new salary of $27.50. If you want to place a towel bar 9 3/4 inches long in the center of a door that is 27 1/2 inches wide, you will need to place the bar about 9 inches from each edge; this estimate can be used as a check on the exact computation.	1, 2
7.EE.4 Use variables to represent quantities in a real-world or mathematical problem, and construct simple equations and inequalities to solve problems by reasoning about the quantities.	3
7.EE.4a Solve word problems leading to equations of the form px + q = r and p(x + q) = r, where p, q, and r are specific rational numbers. Solve equations of these forms fluently. Compare an algebraic solution to an arithmetic solution, identifying the sequence of the operations used in each approach. For example, the perimeter of a rectangle is 54 cm. Its length is 6 cm. What is its width?	3

Continued on next page

Common Core State Standard	***Let's Take Care of the Lo'i* Unit**
7.EE.4b Solve word problems leading to inequalities of the form px + q > r or px + q < r, where p, q, and r are specific rational numbers. Graph the solution set of the inequality and interpret it in the context of the problem. For example: As a salesperson, you are paid $50 per week plus $3 per sale. This week you want your pay to be at least $100. Write an inequality for the number of sales you need to make, and describe the solutions.	3
Geometry	
7.G.1 Solve problems involving scale drawings of geometric figures, including computing actual lengths and areas from a scale drawing and reproducing a scale drawing at a different scale.	4
7.G.2 Draw (freehand, with ruler and protractor, and with technology) geometric shapes with given conditions. Focus on constructing triangles from three measures of angles or sides, noticing when the conditions determine a unique triangle, more than one triangle, or no triangle.	4
7.G.3 Describe the two-dimensional figures that result from slicing three-dimensional figures, as in plane sections of right rectangular prisms and right rectangular pyramids.	4
7.G.4 Know the formulas for the area and circumference of a circle and use them to solve problems; give an informal derivation of the relationship between the circumference and area of a circle.	4
7.G.5 Use facts about supplementary, complementary, vertical, and adjacent angles in a multi-step problem to write and solve simple equations for an unknown angle in a figure.	4
7.G.6 Solve real-world and mathematical problems involving area, volume and surface area of two- and three-dimensional objects composed of triangles, quadrilaterals, polygons, cubes, and right prisms.	4
Statistics and Probability	
7.SP.1 Understand that statistics can be used to gain information about a population by examining a sample of the population; generalizations about a population from a sample are valid only if the sample is representative of that population. Understand that random sampling tends to produce representative samples and support valid inferences.	5

Continued on next page

Common Core State Standard	*Let's Take Care of the Lo'i* Unit
7.SP.2 Use data from a random sample to draw inferences about a population with an unknown characteristic of interest. Generate multiple samples (or simulated samples) of the same size to gauge the variation in estimates or predictions. For example, estimate the mean word length in a book by randomly sampling words from the book; predict the winner of a school election based on randomly sampled survey data. Gauge how far off the estimate or prediction might be.	5
7.SP.3 Informally assess the degree of visual overlap of two numerical data distributions with similar variabilities, measuring the difference between the centers by expressing it as a multiple of a measure of variability. For example, the mean height of players on the basketball team is 10 cm greater than the mean height of players on the soccer team, about twice the variability (mean absolute deviation) on either team; on a dot plot, the separation between the two distributions of heights is noticeable.	5
7.SP.4 Use measures of center and measures of variability for numerical data from random samples to draw informal comparative inferences about two populations. For example, decide whether the words in a chapter of a seventh-grade science book are generally longer than the words in a chapter of a fourth-grade science book.	5
7.SP.5 Understand that the probability of a chance event is a number between 0 and 1 that expresses the likelihood of the event occurring. Larger numbers indicate greater likelihood. A probability near 0 indicates an unlikely event, a probability around 1/2 indicates an event that is neither unlikely nor likely, and a probability near 1 indicates a likely event.	6
7.SP.6 Approximate the probability of a chance event by collecting data on the chance process that produces it and observing its long-run relative frequency, and predict the approximate relative frequency given the probability. For example, when rolling a number cube 600 times, predict that a 3 or 6 would be rolled roughly 200 times, but probably not exactly 200 times.	6
7.SP.7 Develop a probability model and use it to find probabilities of events. Compare probabilities from a model to observed frequencies; if the agreement is not good, explain possible sources of the discrepancy.	6

Continued on next page

Common Core State Standard	***Let's Take Care of the Lo'i* Unit**
7.SP.7a Develop a uniform probability model by assigning equal probability to all outcomes, and use the model to determine probabilities of events. For example, if a student is selected at random from a class, find the probability that Jane will be selected and the probability that a girl will be selected.	6
7.SP.7b Develop a probability model (which may not be uniform) by observing frequencies in data generated from a chance process. For example, find the approximate probability that a spinning penny will land heads up or that a tossed paper cup will land open-end down. Do the outcomes for the spinning penny appear to be equally likely based on the observed frequencies?	6
7.SP.8 Find probabilities of compound events using organized lists, tables, tree diagrams, and simulation.	6
7.SP.8a Understand that, just as with simple events, the probability of a compound event is the fraction of outcomes in the sample space for which the compound event occurs.	6
7.SP.8b Represent sample spaces for compound events using methods such as organized lists, tables and tree diagrams. For an event described in everyday language (e.g., "rolling double sixes"), identify the outcomes in the sample space which compose the event.	6
7.SP.8c Design and use a simulation to generate frequencies for compound events. For example, use random digits as a simulation tool to approximate the answer to the question: If 40% of donors have type A blood, what is the probability that it will take at least 4 donors to find one with type A blood?	6

Unit 1: The Number System

In this unit, we'll learn how to use positive and negative integers, fractions, and decimals to solve problems through exploration of traditional Hawaiian foods and the range of elevation on the Big Island. There are four activities in this unit. *Module 1* involves the use of adding and subtracting integers to help climb a cliff to pick limu. *Module 2* explores temperature changes with different altitudes by multiplying and dividing integers. *Module 3* supports the harvesting of kalo with the help of rational numbers. The final activity is cumulative and incorporates concepts from each of the previous activities in this unit.

Common Core State Standards

Common Core State Standard	Module 1	Module 2	Module 3	Unit 1
7.NS.1 Apply and extend previous understandings of addition and subtraction to add and subtract rational numbers; represent addition and subtraction on a horizontal or vertical number line diagram.	X		X	X
7.NS.1a Describe situations in which opposite quantities combine to make 0. For example, a hydrogen atom has 0 charge because its two constituents are oppositely charged.	X		X	X
7.NS.1b Understand p + q as the number located a distance |q| from p, in the positive or negative direction depending on whether q is positive or negative. Show that a number and its opposite have a sum of 0 (are additive inverses). Interpret sums of rational numbers by describing real-world contexts.	X		X	X
7.NS.1c Understand subtraction of rational numbers as adding the additive inverse, p − q = p + (−q). Show that the distance between two rational numbers on the number line is the absolute value of their difference, and apply this principle in real-world contexts.	X		X	X
7.NS.1d Apply properties of operations as strategies to add and subtract rational numbers.	X		X	X
7.NS.2 Apply and extend previous understandings of multiplication and division and of fractions to multiply and divide rational numbers.		X	X	X
7.NS.2a Understand that multiplication is extended from fractions to rational numbers by requiring that operations continue to satisfy the properties of operations, particularly the distributive property, leading to products such as (−1)(−1) = 1 and the rules for multiplying signed numbers. Interpret products of rational numbers by describing real-world contexts.		X	X	X
7.NS.2b Understand that integers can be divided, provided that the divisor is not zero, and every quotient of integers (with non-zero divisor) is a rational number. If p and q are integers, then −(p/q) = (−p)/q = p/(−q). Interpret quotients of rational numbers by describing real-world contexts.		X	X	X

Continued on next page

Common Core State Standard	Module 1	Module 2	Module 3	Unit 1
7.NS.2c Apply properties of operations as strategies to multiply and divide rational numbers.		X	X	X
7.NS.3 Solve real-world and mathematical problems involving the four operations with rational numbers. (Computations with rational numbers extend the rules for manipulating fractions to complex fractions.)	X	X	X	X
7.EE.3 Solve multi-step real-life and mathematical problems posed with positive and negative rational numbers in any form (whole numbers, fractions, and decimals), using tools strategically. Apply properties of operations as strategies to calculate with numbers in any form; convert between forms as appropriate; and assess the reasonableness of answers using mental computation and estimation strategies. For example: If a woman making $25 an hour gets a 10% raise, she will make an additional 1/10 of her salary an hour, or $2.50, for a new salary of $27.50. If you want to place a towel bar 9 3/4 inches long in the center of a door that is 27 1/2 inches wide, you will need to place the bar about 9 inches from each edge; this estimate can be used as a check on the exact computation.	X	X	X	X

Module 1: Adding and Subtracting Integers *Activity*

Limu (seaweed) is a very important part of Hawaiian life. It is often used for food, decoration, and important ceremonies. Kohu, ʻeleʻele, and līpoa are some of the most delicious types of limu to eat. Limu huluhuluwaena was Queen Liliʻoukalani favorite limu. She loved it so much that she had it brought in from Maui to plant in Oʻahu.

Poke bowls are more delicious with limu!

In this activity, we will carefully climb up and down a rock wall to collect some limu.

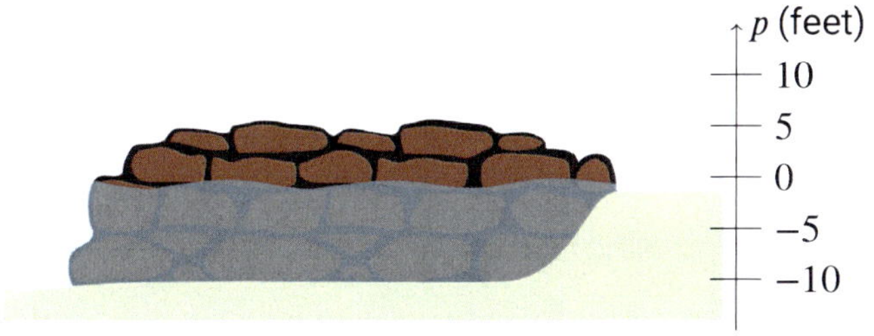

1. You are starting at a position of $p = 0$ feet. Below are some descriptions of where you should climb to find limu. Write your position, p, after each description. The first two are done for you.

   (a) Climb up 2 feet from where you started to get the first limu.
   $p = 0 + 2 = 2$

   (b) Now climb down 3 feet to get to the next piece of limu.
   $p = 2 - 3 = -1$

   (c) The next piece of limu can be found if you decrease your position by 6 feet. Where is it located?

   (d) Add 2 feet to your position to get the next limu. What is your position now?

   (e) You are too deep in the water to get the next limu so let's go to a position that is "less deep" by subtracting -3 feet. What is your new position?

   (f) The final limu requires you to go deeper. So let's go "more deep" by adding -8.5 feet to get to the last limu. What is your final position?

2. Were there any positions that were opposite numbers? If so, what were these numbers?

3. Plot and label the positions where you found the six pieces of limu. Label the limu **a-f** based on your answers above. The first two are done for you.

4. Use the number line to help you answer the following questions.
   (a) How many feet apart are the first limu and the last limu?

   (b) How many feet apart are the first limu, and the limu in part 1e?

   (c) What if you wanted to return to where you started $p = 0$ after grabbing the last limu? What number do you need to add to get to $p = 0$?

   (d) Maybe you could have saved time by starting at the bottom of the rock wall and collecting limu on the way up. Using the letters a-f, order the limu from lowest to highest.

5. Share a time when you have eaten something that you found outside.

Module 2: Multiplying and Dividing Integers *Activity*

Hawai'i Island has 8 of the 13 possible climate zones. Ranging from tropical beaches to dry deserts to snow covered mountains, it is a truly beautiful and unique place. Mauna Kea is the tallest mountain in Hawai'i at $13,796$ feet (ft) above the sea level.

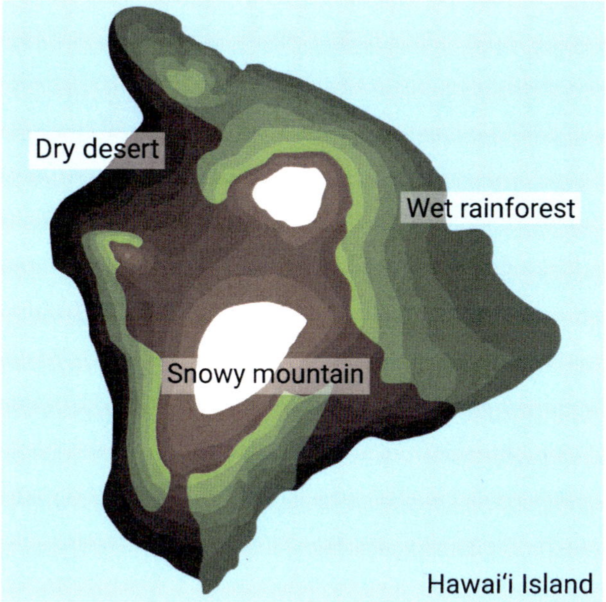

In an ahupua'a on Hawai'i, there is a temperature change of about -3°F for every $1,000$ ft increase in elevation.

1. If you went up 1 foot in elevation, what would be the temperature change? Make sure to include a negative sign if your answer is negative. Is your answer a rational number? How do you know?

2. Imagine it is 84°F at $3,000$ ft above sea level. Fill in the chart to help you answer the following questions. You do NOT need to use your previous answer to fill out this table.

Temperature	Elevation
	$2,000$ ft.
84°F	$3,000$ ft.
81°F	
	$5,000$ ft.
	$6,000$ ft.

3. If you wanted to cool off, would you go up or down the mountain? Please explain your reasoning.

4. What temperature would you expect at sea level?

5. What temperature would you expect at the top of Mauna Kea? Round the elevation of Mauna Kea to the nearest thousand to estimate your answer. Explain the reasoning for your answer.

6. Mauna Kea actually starts on the sea floor, which means that most of the mountain is underwater. If it started at sea level, it would be over $33,000$ feet tall! That's taller than the tallest mountain in the world, Mount Everest! What temperature would you expect at $33,000$ feet?

7. Kalo (taro) plants typically thrive in temperature between 78°F and 95°F. If you were to build a lo'i (taro garden), what elevation would you make it at? Please explain your reasoning.

Module 3: Rational Numbers *Activity*

The kalo plant has been very important to the people in Hawai'i for a long time. Whether it is for a graduation party, a family dinner, or a religious offering, the kalo is still a staple for Hawaiians. So why don't more people plant it? Well, kalo is difficult to grow and reproduce after it is harvested. Most plants will give you hundreds or thousands of seeds that might grow into new plants, but not kalo. Instead, to regrow kalo, you must start with a healthy plant. Next, you'll cut off and replant a small part of it. Usually, you have to cut off the huli, or the top part of the kalo where the leaves begin. Sometimes, if you are lucky, you also get a 'ohā, which is a like a baby kalo that grows out of a bigger kalo. By planting a huli or a 'ohā and taking care of it for about a year, you might end up with a new kalo.

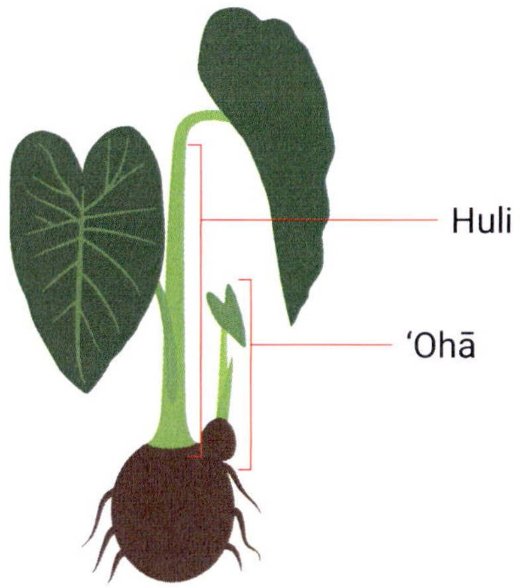

The parts of a kalo that can be replanted

Uncle Ikaika has a large farm, but he is only using 0.5 acres of it to grow kalo. His nephew has a graduation party next year, and he wants to grow more kalo for the party.

1. Uncle Ikaika needs to plant more kalo, so he uses all the huli that he has saved to grow $1/10$ acres of more kalo. How many acres of kalo does Uncle Ikaika have now?

2. Uncle Ikaika's neighbor has a much bigger farm, and he wanted to help by donating some of his 'ohā to Uncle Ikaika's. With these donations, Uncle Ikaika is able to grow 35% more kalo than he had in part 1! How many acres of kalo does Uncle Ikaika have now?

3. It's a good thing that he planted so much kalo! At the party, $8/9$ of the kalo was eaten. How many acres of kalo was eaten, and how many acres of kalo does Uncle Ikaika have left?

Unit 1: *Cumulative Activity*

We want to take care of a 120 thousand square feet lo'i in Kanē'ohe, a rainy part of O'ahu. Unfortunately, two major floods have damaged the lo'i.

1. The first flood destroyed **5/16**th of the lo'i. Below is a grid of 120 squares representing 120 thousand square feet of the lo'i. Color/fill in the squares to represent how many thousand square feet of the lo'i was destroyed after the first flood.

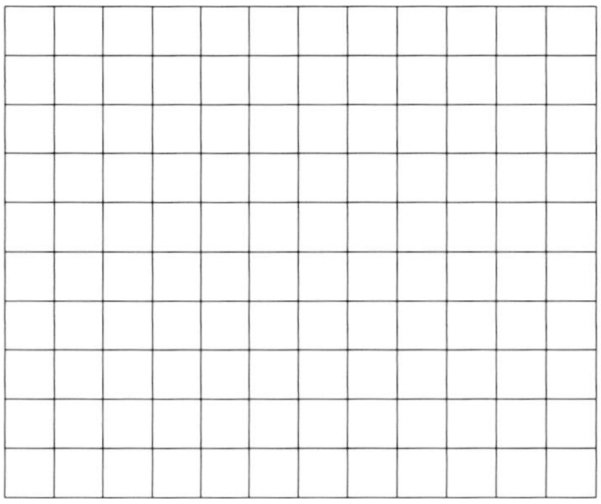

2. The second flood destroyed another **22.5 thousand square feet** of the lo'i. Below is a grid of 120 squares representing 120 thousand square feet of the lo'i. Color/fill in the squares to represent how many thousand square feet of lo'i was destroyed after both floods.

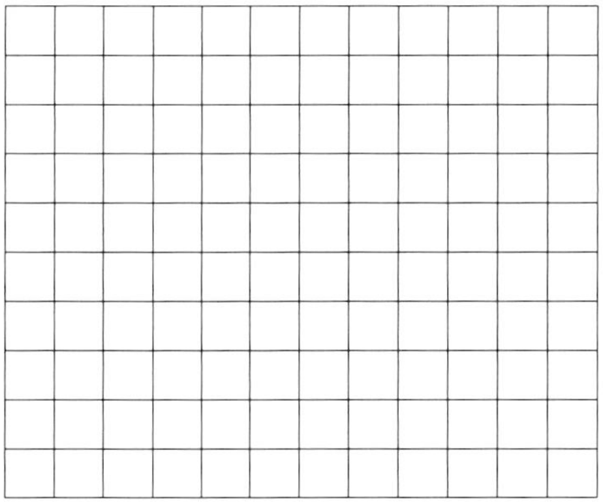

3. What is the size of the lo'i after the two floods? Give your answer in thousands of square feet.

4. You will need a lot of help to take care of the remaining parts of the lo'i. You decide to look for volunteers from your community and your uncle helps you figure out how many volunteers are needed. His tip is to use the **area of the lo'i** (in square feet), and **divide** it by the fraction **25,000/30**. Based on your answer in part 2 and your uncle's rule of thumb, how many volunteers do you need? Note that the area is in square feet, not thousands of square feet.

5. Your uncle's rule might be more complicated than it needs to be. Can you and your partner come up with four (4) other rules that mean the exact same thing?

6. Come up with some ideas on how we might be able to protect a lo'i from severe rain. Share your ideas with a partner or in the online comment section.

Unit 2: Ratios and Proportional Relationships

In this unit, we'll learn how to recognize and describe proportional relationships and use ratios and rates to solve problems through farming lo'i kalo and u'ala (sweet potato). There are three activities in this unit. *Module 4* involves using ratios and proportionality to explore how farmers build Lo'i. *Module 5* focuses on evaluating how a 'uala grows through the seasons with the help of proportions and percents. The final activity is cumulative and incorporates concepts from each of the previous activities in this unit.

Common Core State Standards

Common Core State Standard	Module 4	Module 5	Unit 2
7.RP.1 Compute unit rates associated with ratios of fractions, including ratios of lengths, areas and other quantities measured in like or different units. For example, If a person walks 1/2 mile in each 1/4 hour, compute the unit rate as the complex fraction (1/2)/(1/4) miles per hour, equivalently 2 miles per hour.	X		X
7.RP.2 Recognize and represent proportional relationships between quantities.	X		X
7.RP.2a Decide whether two quantities are in a proportional relationship, e.g., by testing for equivalent ratios in a table or graphing on a coordinate plane and observing whether the graph is a straight line through the origin.	X		X
7.RP.2b Identify the constant of proportionality (unit rate) in tables, graphs, equations, diagrams, and verbal descriptions of proportional relationships.	X		X
7.RP.2d Explain what a point (x, y) on the graph of a proportional relationship means in terms of the situation, with special attention to the points (0, 0) and (1, r) where r is the unit rate.	X		X
7.RP.3 Use proportional relationships to solve multistep ratio and percent problems. Examples: simple interest, tax, markups and markdowns, gratuities and commissions, fees, percent increase and decrease, percent error.		X	X
7.EE.2 Understand that rewriting an expression in different forms in a problem context can shed light on the problem and how the quantities in it are related. For example, a + 0.05a = 1.05a means that "increase by 5%" is the same as "multiply by 1.05."		X	X

Continued on next page

Common Core State Standard	Module 4	Module 5	Unit 2
7.EE.3 Solve multi-step real-life and mathematical problems posed with positive and negative rational numbers in any form (whole numbers, fractions, and decimals), using tools strategically. Apply properties of operations as strategies to calculate with numbers in any form; convert between forms as appropriate; and assess the reasonableness of answers using mental computation and estimation strategies. For example: If a woman making \$25 an hour gets a 10% raise, she will make an additional 1/10 of her salary an hour, or \$2.50, for a new salary of \$27.50. If you want to place a towel bar 9 3/4 inches long in the center of a door that is 27 1/2 inches wide, you will need to place the bar about 9 inches from each edge; this estimate can be used as a check on the exact computation.		X	X

Module 4: Ratios and Proportionality *Activity*

Turning undisturbed ground into a kalo lo'i takes a lot of work.

First, you have to pull all of the weeds out. Then, you need to stomp until the unwanted plants die, and the dirt turns into mud from the moisture in the ground. Then, you have to push the mud into mounds. If you do it correctly, the space around the mound will start to pick up water. Only then can you begin planting your kalo.

Building a kalo mound in a lo'i

The amount of time it takes to build a lo'i often depends on how skilled the farmer is. However, it can also depend on the location of the lo'i. Let's take a look at a few farmers who are building a lo'i from scratch. We will pay attention to rate that they make a mound for the kalo.

Farmer	**Kawika**
Location	Kapolei
Lo'i building	Kawika can build 10 mounds every 4 hours.

Farmer	**Kalani**
Location	Waiau
Lo'i building	After t hours, Kalani builds $3.5t$ mounds

Farmer	**Kekoa**
Location	Waimānalo
Lo'i building	(table below)

Hours	Mounds built
0	0
1	8
2	12
3	14
4	15
5	15.5

Farmer	**Mina**
Location	Mililani
Lo'i building	(table below)

Hours	Mounds built
0	0
2	6
4	12
6	18

Farmer	**Kami**
Location	Wai'anae
Lo'i building	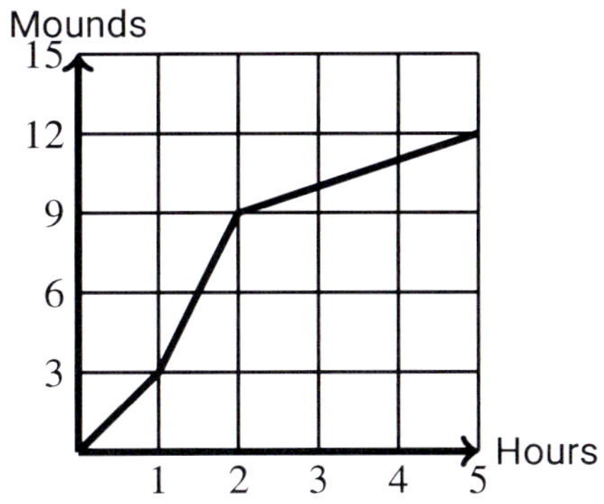

Farmer	**Lani**
Location	Kāne'ohe
Lo'i building	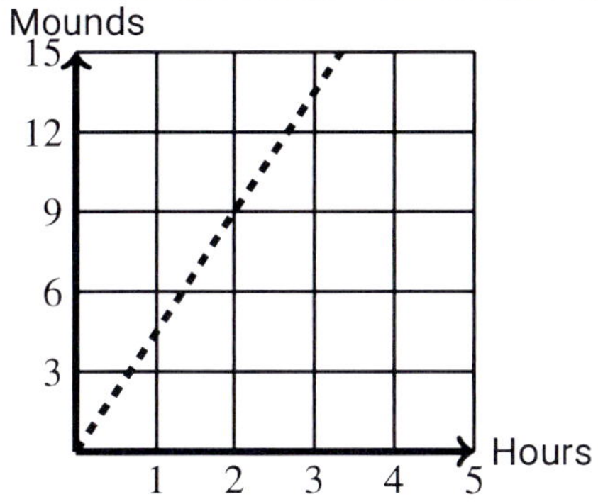

1. Most of the farmers were planting at a constant rate. However, the information for **two** of these farmers did not show a *proportional relationship*. Which two farmers were **NOT** proportional? How can you tell? If you are not sure, work with a partner before moving on.

2. For the other four farmers, what is the unit rate of mounds made per hour? It is okay to answer as a decimal or a simplified fraction.

3. Graph and label the data for all six farmers. Two of them have already been done for you. Make sure to draw lines, especially when you are given data from a table. Make sure that your lines do not stop before reaching the edges of the graph.

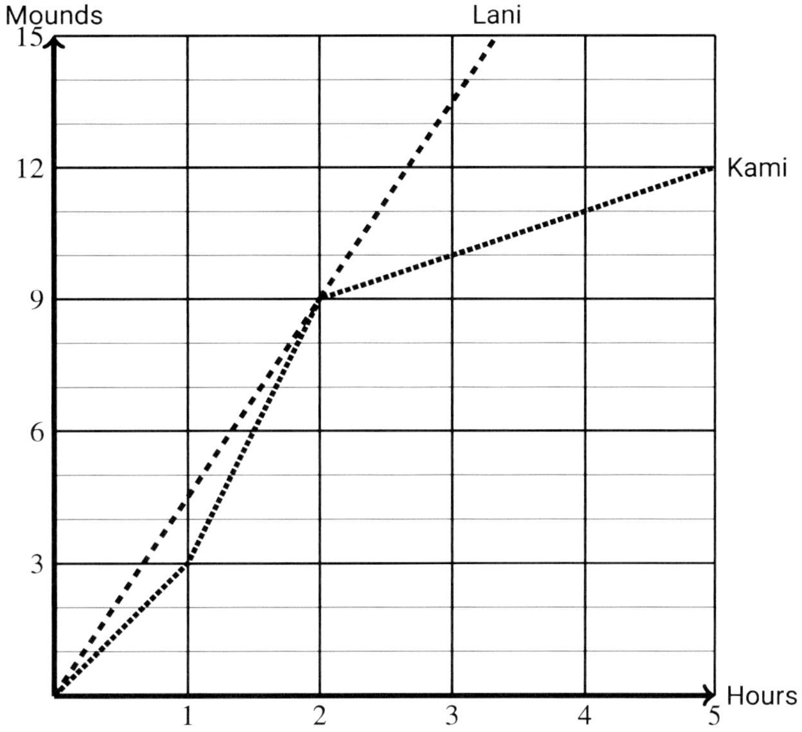

4. Rank the farmers from highest to lowest number of mounds made after 5 hours.

5. Why do you think some environments are harder to build a loʻi on than others? Share and discuss your thoughts with your partner or in the online comment section.

Module 5: Proportions and Percent *Activity*

The ʻuala or sweet potato is a very important vegetable in Hawaiʻi. It grows quickly and easily with many uses such as food, medicine, and decoration.

ʻUala

Suppose that you had grown a big batch of ʻuala in your garden that is about 125 square feet in area.

1. This winter, your ʻuala batch is thriving, and its area has grown 20% bigger. How big is your ʻuala batch at the end of winter?

2. The following summer was unusually hot, and 20% of the ʻuala died. You told your neighbor that you now have 120 square feet of ʻuala, but he did not believe you. He said "if your 125 square feet batch of ʻuala got bigger by 20% then, smaller by 20%, you should be back to 125 square feet." Can you explain why you are right and your neighbor is wrong? Be sure to calculate how much ʻuala you have now, and show your work.

In some villages on Hawai'i Island, the 'uala is considered to be very beautiful. Traditionally, the villagers would gather their best 'uala and decorate the side of the road when a neighboring chief came to visit. Before the chief arrived, the villagers would put the 'uala on the right side of the road. Then, they would move it to the other side of the road so the 'uala would still be on the right side when the chief left.

3. A chief is visiting your village. You decide to give 21 square feet of your best 'uala that survived the summer to decorate his arrival. What percent of your 120 square feet 'uala did you give away?

4. Of all the plants that you eat, which one do you think is the prettiest and why? Please share with your partner or in the online comment section.

Unit 2: *Cumulative Activity*

In one year, a typical 12 feet by 12 feet lo'i patch can produce 50 kalo.

1. Assuming that this rate is proportional, write the rate of kalo to square feet as a *simplified* fraction.

$$\frac{\rule{3cm}{0.4pt} \text{ kalo}}{\text{square feet}}$$

2. Find the total area of 5 lo'i patches. Remember each lo'i patch is 12 by 12 feet.

3. How many kalo can be produced by 5 lo'i patches?

4. How many kalo can a 12 by 18 feet patch produce in one (1) year?

5. A farmer named Kaliko has a lo'i patch that is 16 feet wide by 18 feet long. How many kalo can he typically produce in one (1) year?

6. Kaliko wants to break apart and rebuild his lo'i patch from part 5. He wants to produce 25% more kalo. What is the area of new lo'i that he needs?

7. Work with a partner and draw at least three possible lo'i that have the area needed in part 6. Be sure to label the lengths of each side.

8. Most of us have no idea how the food we eat made it to our tables. Describe a fruit or vegetable that you enjoy eating or have eaten recently. Now, imagine and share a story of how this plant might have gotten to your dinner table. Talk about where you think it was planted, who planted it, how it got to you, etc. Share with your partner or in the online comment section.

Unit 3: Expressions, Equations, and Inequalities

In this unit, we'll learn how to write and use algebraic equations and inequalities to solve problems through volunteering at the loʻi, planting sugar cane, and removing invasive fish species. There are four activities in this unit. *Module 6* involves evaluating loʻi volunteers' work hours through expressions and equations. *Module 7* focuses on using inequalities for planning a kō plantation. There are two cumulative activities in this unit. Each of the cumulative activities incorporate concepts from each of the previous activities in this unit.

Common Core State Standards

Common Core State Standard	Module 6	Module 7	Unit 3
7.EE.1 Apply properties of operations as strategies to add, subtract, factor, and expand linear expressions with rational coefficients.	X		X
7.EE.2 Understand that rewriting an expression in different forms in a problem context can shed light on the problem and how the quantities in it are related. For example, a + 0.05a = 1.05a means that "increase by 5%" is the same as "multiply by 1.05."			X
7.EE.4 Use variables to represent quantities in a real-world or mathematical problem, and construct simple equations and inequalities to solve problems by reasoning about the quantities.	X	X	X
7.EE.4a Solve word problems leading to equations of the form px + q = r and p(x + q) = r, where p, q, and r are specific rational numbers. Solve equations of these forms fluently. Compare an algebraic solution to an arithmetic solution, identifying the sequence of the operations used in each approach. For example, the perimeter of a rectangle is 54 cm. Its length is 6 cm. What is its width?	X	X	X
7.EE.4b Solve word problems leading to inequalities of the form px + q > r or px + q < r, where p, q, and r are specific rational numbers. Graph the solution set of the inequality and interpret it in the context of the problem. For example: As a salesperson, you are paid $50 per week plus $3 per sale. This week you want your pay to be at least $100. Write an inequality for the number of sales you need to make, and describe the solutions.		X	X

Module 6: Expressions and Equations *Activity*

A high school on O'ahu requires their seniors to volunteer in the Hawaiian community. Four senior high school friends decide to volunteer at a lo'i in Kāne'ohe (O'ahu).

Hiapo is the leader of the volunteer group. First, she works 6 hours in the lo'i. She really likes it, and she decides to work 4 hours each day after that. **Kalua** always works half as many hours as Hiapo and **Kekolu** works a third as many hours as Hiapo. **Maka** works 9 hours on his first day, but he didn't like it as much. So, he decided to work just 2 hours every 3 days after that. This is equivalent to working $2/3$ of an hour per day.

1. Complete the following table summarizing the four volunteers.

Volunteer	Starting hours	Hours per day	Total hours volunteered t days after the starting day
Hiapo	6	4	$6 + 4t$
Kalua		2	
Kekolu	2		
Maka			

2. The school is asking this group to volunteer for a total of 100 hours together.

   (a) Find an algebraic expression showing the total number of hours worked by this group t days after the starting day.

   (b) Use your expression from part 2a to find how many days it takes for this group to reach a total of 100 volunteer hours.

3. To prevent students from slacking, the school also requires that each student volunteer for at least 20 hours. Using the answer for t that you found in 2b, check if all four students were able to complete over 20 hours each. Show your work below.

4. Do you do volunteer work or community service? If so, what is it and why do you do it? If you aren't already helping your community, what is a community project that you would be interested in and why? Share with a partner or in the online comment section.

Module 7: Inequalities *Activity*

Kō, or sugar cane, is a giant grass that made life in Ancient Hawai'i more enjoyable. It is still a delicious snack that is easy to find and bring on long journeys. The kō makes foods like haupia and kūlolo much sweeter, and it protects the lo'i by holding the soil together during heavy rain and strong winds.

Kō

A group of farmers are planning to plant kō around a lo'i, but first, they must pull out all the weeds in the area. Right now there are 5000 square feet of weeds that need to be replaced by kō. One team of farmers will pull out weeds and, the other team will prepare and plant the new kō.

1. The weed-removal team is able to remove 20 square feet of weeds every 3 hours. Write the expression for the **amount of weeds remaining** (in square feet) after pulling them for t hours. Remember that the farmers start with 5000 square feet of weeds.

2. The kō-planting team is able to plant 5.5 kō every hour. Write the expression for the **amount of kō planted** (in square feet) after planting for t hours. Remember that the farmers start with 0 square feet of kō.

3. For each of the following statements, **write the equation or inequality** it describes, then **solve the equation or inequality**.

   (a) The area of **weeds remaining** after t hours of removal is 3000 square feet.

   (b) The area of **weeds remaining** after t hours of removal is less than 1000 square feet.

   (c) The amount of **kō planted** after t hours of planting is 1650 square feet.

   (d) The amount of **kō planted** after t hours of planting is greater than or equal to 3300 square feet.

Unit 3: *Cumulative Activity 1*

The roi fish, also known as the peacock grouper, is a very invasive fish in Hawai'i. In fact, many scientists believe that it is now the top predator along Hawai'i's shorelines. It is estimated that in a three-square-mile area off the Kona Coast, the roi will eat about 8.2 million fish per year.

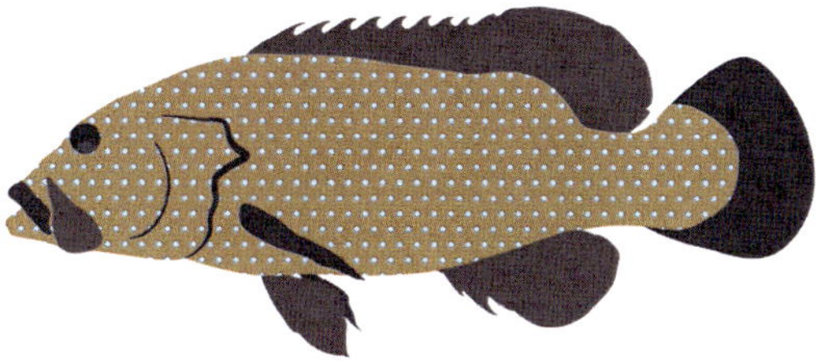

Roi fish

Many local fishermen despise the roi because they are a major threat to the local fish population. Fishermen are also afraid to eat roi because their meat has a reputation of making people sick. To diminish the population of the invasive roi, the State of Hawai'i has been arranging spearfishing events called the Roi Roundup.

Keoni and his dad Kawika joined this year's Roi Roundup. They arrive early to the event because they want to do their part in helping the native environment recover.

1. At $4{:}00$ AM, before the Roi Roundup started, Keoni catches 3 pounds of roi. Every hour after $4{:}00$ AM, Keoni catches $\frac{5}{3}$ pounds of roi. Write an expression for the amount of fish caught by Keoni t hours after $4{:}00$ AM.

2. At $4{:}00$ AM, Kawika catches 5 pounds of roi. Every hour after $4{:}00$ AM, Kawika catches $\frac{3}{2}$ pounds of roi. Write an expression for the amount of fish caught by Kawika t hours after $4{:}00$ AM.

3. After how many hours will Keoni and Kawika have the same amount of roi?

4. Write an expression for the total amount of roi caught by Keoni and Kawika t hours after 4:00 AM.

5. Last year, Keoni and Kawika spearfished until 4:00 PM ($t = 12$) and caught $33\frac{1}{3}$ pounds of fish. This year, they are catching even more. How many hours after 4:00 AM will Keoni and Kawika need to catch at least $33\frac{1}{3}$ pounds of roi?

6. Work with a partner to discuss why people are only allowed to spearfish in this event. Why don't they allow other ways of catching fish like using a net or a fishing pole? What are the benefits of only using a spear?

7. With your partner or in the online comment section, share your favorite seafood. Please explain whether or not it is something that is found in the waters of Hawai'i. If it is found in Hawai'i, is it originally from here or was it brought here by people? Feel free to look online for research and pictures (with your teacher's permission).

Unit 3: *Cumulative Activity 2*

Many characters in traditional Hawaiian stories have super powers. For example, Kūʻulakai (also known as Kūʻula) was known to build large fishponds by himself and communicate with the fish in the sea. According to Hawaiian legend, he would often call fish in from the sea to help those in need of fish or call fish away from entire coastlines to punish evil doers.

Let's imagine that Kūʻula is planning to build a rectangular fishpond. One side of the fishpond will be part of the shoreline and the other three sides will be stone walls sticking out into the ocean. He could measure the three stone walls by swimming in the ocean.

Rectangular fishpond

1. Kūʻula measures the length of his first wall by swimming straight out into the ocean. He takes 8 strokes and then swims 3 more feet. If s is the length of his swim stroke (in feet), write the expression for length of the first wall.

2. Next, Kūʻula makes a sharp $90°$ turn and swims along the shoreline. He takes 10 strokes to measure his second wall. If s is the length of his swim stroke (in feet), write the expression for length of the second wall.

3. Finally, he swims straight back to shore, traveling the same distance that he swam out in the beginning. So, the measurement of his last wall is the same as his first. Write the expression for the total length of all three stone walls.

4. The three stone walls will have a total length of 240 feet. How long is Kūʻula's swim stroke, s, in feet?

Imagine that a young girl saw the fishpond and wanted to catch a fish. She didn't know how to fish, but after many hours of trying, she finally caught a big one. She's hungry and tired so she carried her fish home. On her way home, she passed by an old man who said that he was really hungry. Without hesitation, the young girl gave away her big fish and went back to the pond, hoping to catch another one before it gets dark.

5. Kū'ula sees this and is impressed by the girl's kindness. He knows the girl is tired and it took a long time for her to catch the fish, so he calls out for the fish to come to his pond so that the girl could easily catch a new fish. The amount of fish in the pond increases by 140% as a result. If f was the amount of fish that were in the pond earlier, in pounds, write an expression for the amount of fish that are in the pond now.

6. More than 240 pounds of fish are in this pond now and the young girl catches many fish with ease! Write this as an **inequality** with the expression from the part 5.

7. At least how many pounds of fish, f, used to be in the pond before Kū'ula called for more?

8. If you could talk to an animal, what kind of animal would it be and why? Share with a partner or in the online comment section.

Unit 4: Geometry

In this unit, we'll learn about scale drawings, and two- and three-dimensional objects composed of triangles, quadrilaterals, polygons, and right prisms through Hawaiian quilt making, constructing fish traps, and building an imu. There are four activities in this unit. *Module 8* involves modeling geometric figures by measuring and drawing a kalo quilt. *Module 9* has two activities that focus on circumference, area, and volume while exploring the size and structure of umu. The final activity is cumulative and incorporates concepts from each of the previous activities in this unit.

For some of the activities in this unit, students will need a ruler with centimeters.

Common Core State Standards

Common Core State Standard	Module 8	Module 9	Unit 4
7.G.1 Solve problems involving scale drawings of geometric figures, including computing actual lengths and areas from a scale drawing and reproducing a scale drawing at a different scale.	X		
7.G.2 Draw (freehand, with ruler and protractor, and with technology) geometric shapes with given conditions. Focus on constructing triangles from three measures of angles or sides, noticing when the conditions determine a unique triangle, more than one triangle, or no triangle.	X		
7.G.3 Describe the two-dimensional figures that result from slicing three-dimensional figures, as in plane sections of right rectangular prisms and right rectangular pyramids.		X	
7.G.4 Know the formulas for the area and circumference of a circle and use them to solve problems; give an informal derivation of the relationship between the circumference and area of a circle.		X	X
7.G.5 Use facts about supplementary, complementary, vertical, and adjacent angles in a multi-step problem to write and solve simple equations for an unknown angle in a figure.	X		
7.G.6 Solve real-world and mathematical problems involving area, volume and surface area of two- and three-dimensional objects composed of triangles, quadrilaterals, polygons, cubes, and right prisms.		X	X

Module 8: Modeling Geometric Figures *Activity*

For this activity, you will need a ruler with centimeters.

As you know, kalo was a very important plant to the Native Hawaiians. It even made its way into the patterns of Hawaiian quilts. Below is an image of a quilt with some lines showing the symmetry of the design. This particular **square** quilt measures 8 feet by 8 feet in real life.

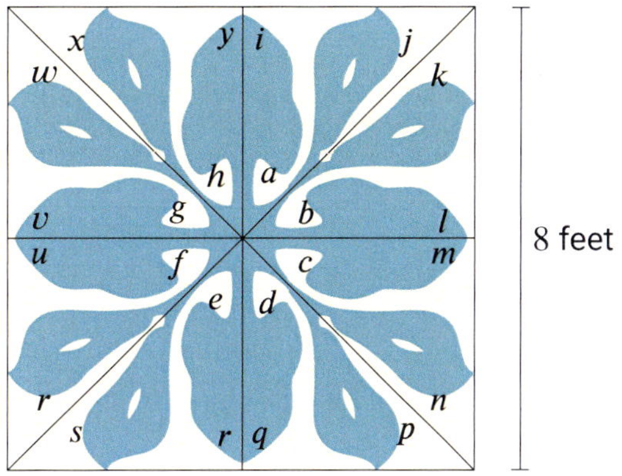

Hawaiian quilt

1. Use a ruler to sketch a scale drawing of this quilt. For your scale, 2.5 centimeters in your drawing should be 2 feet in real life. (You don't have to draw the pattern perfectly, a rough sketch is fine.)

2. Find **2** angles that are adjacent to $\angle a$.

   (a) $\angle$ ____

   (b) $\angle$ ____

3. Find **3 pairs** of vertical angles on the quilt.

   (a) $\angle$ ____ and $\angle$ ____
   (b) $\angle$ ____ and $\angle$ ____
   (c) $\angle$ ____ and $\angle$ ____

4. Find **2 pairs** of complementary angles.

   (a) $\angle$ ____ and $\angle$ ____
   (b) $\angle$ ____ and $\angle$ ____

5. Find **2 pairs** of supplementary angles.

   (a) $\angle$ ____ and $\angle$ ____
   (b) $\angle$ ____ and $\angle$ ____

6. Looking at the original image, it is divided into 8 isosceles triangles. Each triangle has two sides with the same length of 4 feet, and a third side. For any triangle where two sides have a length of 4 what is the shortest and longest possible length for the third side? Explain your reasoning and sketch a drawing to help with your explanation.

Module 9: Circumference, Area, and Volume *Activity 1*

The umu is a fish trap made out of rocks or coral that is, basically, a man-made home for fish. The umu is a place for fish to find food and protection. When a big enough fish begins to get comfortable in its man-made environment, fishermen can surround the umu with a net to catch the big buggah.

Fishing with an umu

We built an umu that is the shape of a trapezoidal prism.

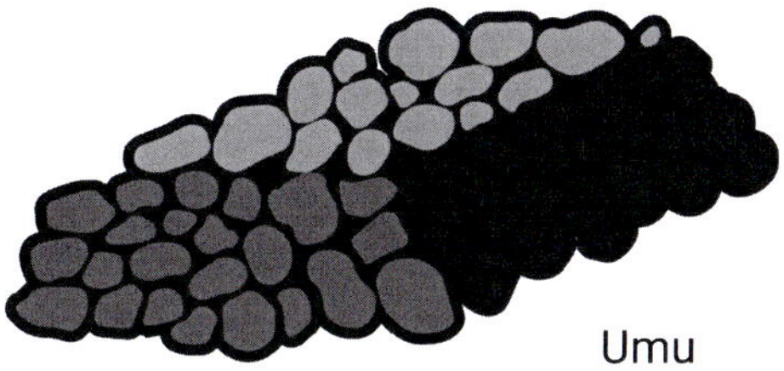

Umu

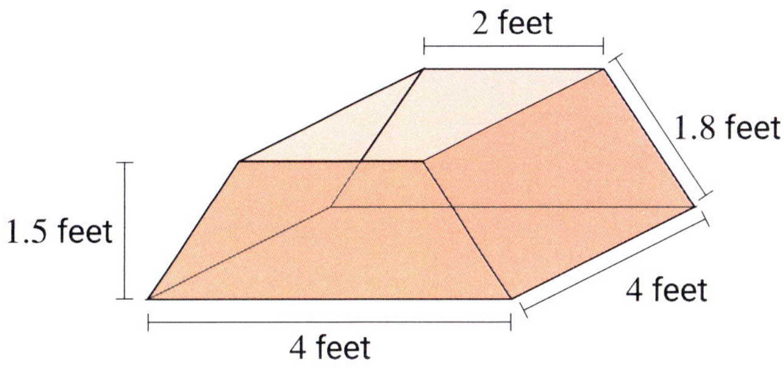

1. Find the volume of this umu. Give your answer in cubic feet.

2. After a few weeks, plenty of fish had swam into the umu. We found that for every cubic foot of umu, there were 0.3 pounds of fish. Find the total weight of the fish in the umu.

There are a few ways to catch the fish in the umu. If you are working with several friends, you can hold a long net and create a fence around the whole umu. Then, someone can take apart the umu and guide the fish into the net. If you are working with one friend, you can make two openings in the umu. You can hold the net at one opening, while your friend scares the fish from the other opening.

3. You are trying to get the fish on your own, so you decide to cover the entire umu with a large net. Then, one by one, you remove the rocks from the umu until the fish have nowhere left to hide. Find the **total** surface area of the top and the four sides of the umu. (Do not include the bottom in your total.)

4. Below are the sizes of some of the circular nets. Cross out the nets that **do not have enough area** to cover the top and four sides of the umu.

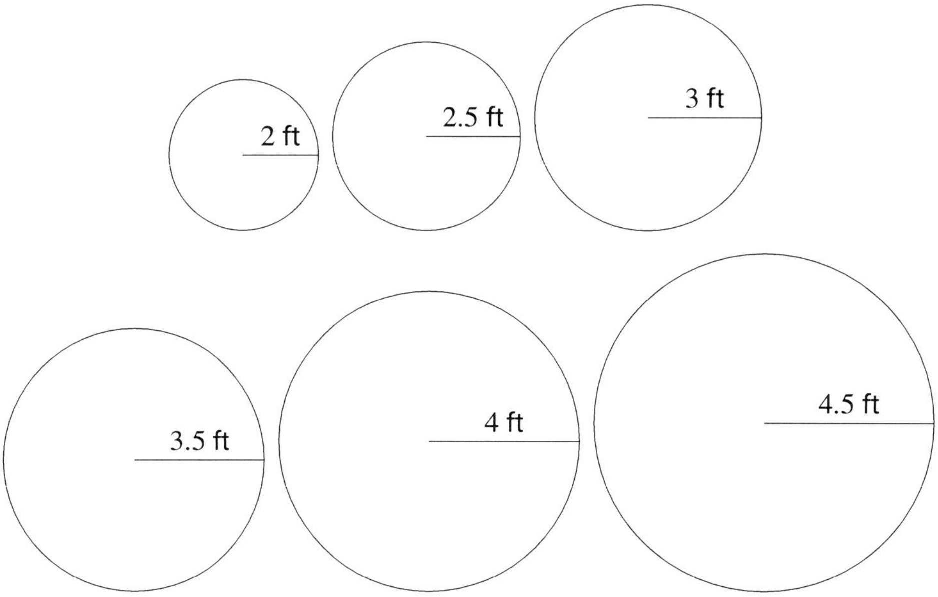

Module 9: Circumference, Area, and Volume *Activity 2*

The umu is a fish trap made out of rocks or coral that is basically, a man-made home for fish. The umu is a place for fish to find food and protection. When a big enough fish begins to get comfortable in its man-made environment, fisherman can surround the umu with a net to catch the big buggah.

Fishing with an umu

We were able to make a cube-shaped umu before a part of it broke off in the waves.

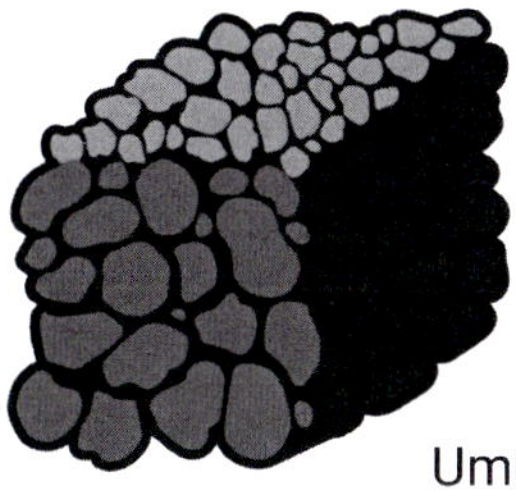

Umu

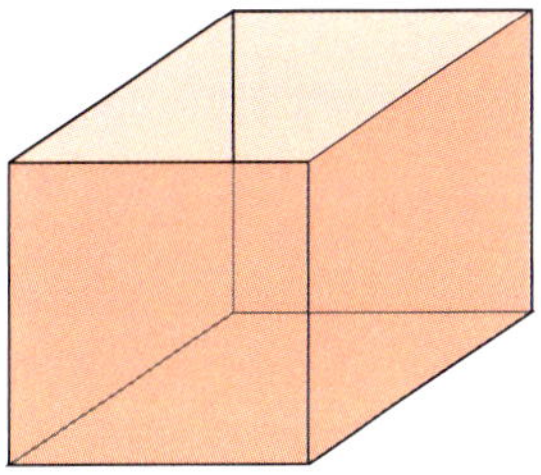

If the top front edge broke off, we would see a **rectangular cross section**.

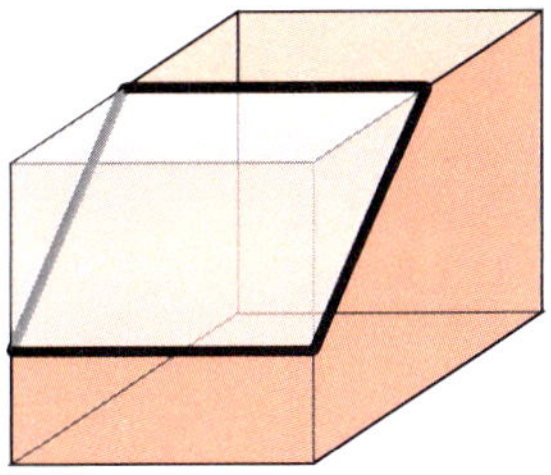

For each of the following questions, three cubes are given to you. **You only have to use one** and may use the other two for practice.

1. How can a piece be broken off and leave behind a **rectangular cross section** that is different from the one shown above? Use four lines to draw this rectangle.

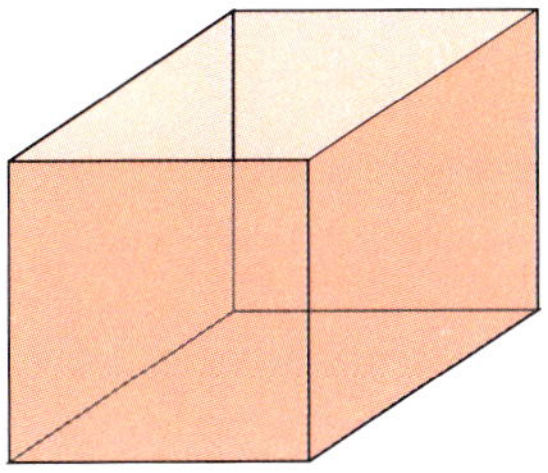
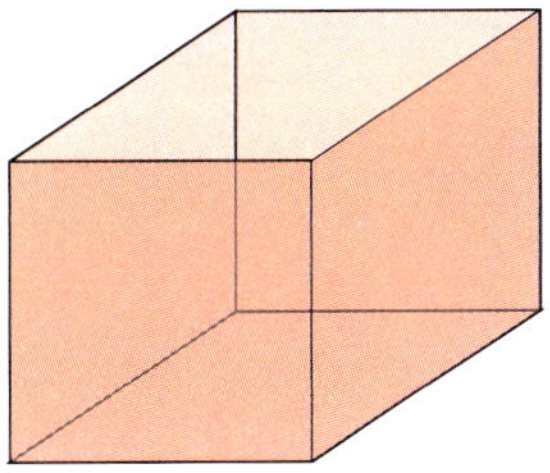
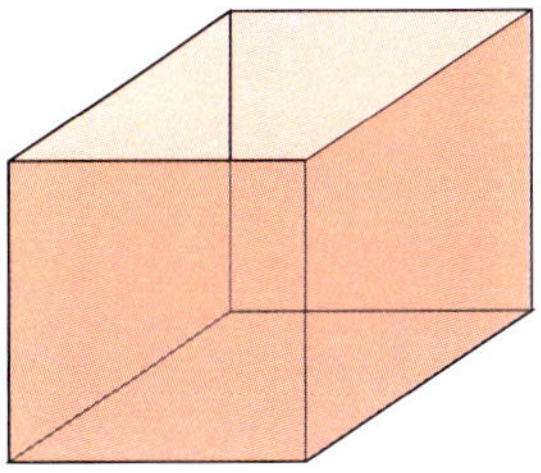

2. How can a piece be broken off and leave behind a **triangular cross section**? Use three lines to draw this triangle.

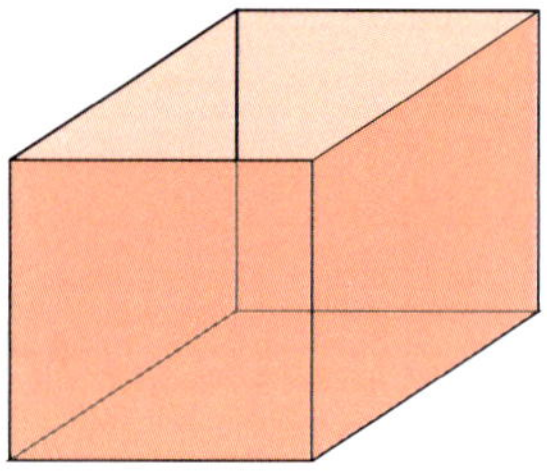 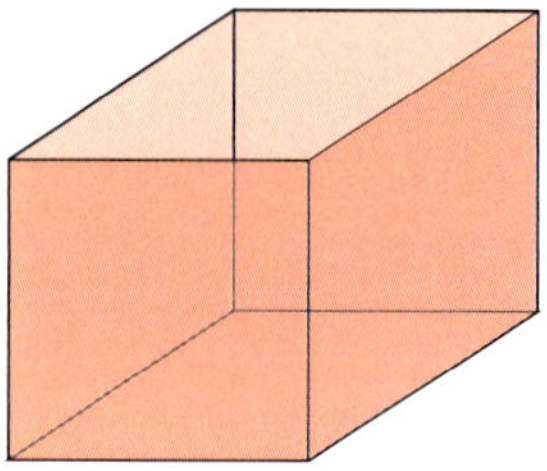 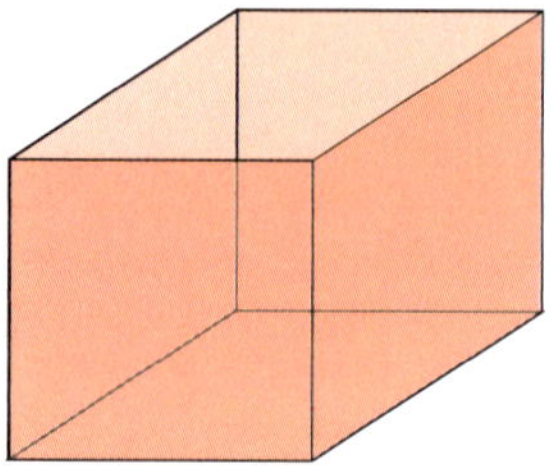

3. How can a piece be broken off and leave behind a **pentagonal cross section**? Use five lines to draw this pentagon.

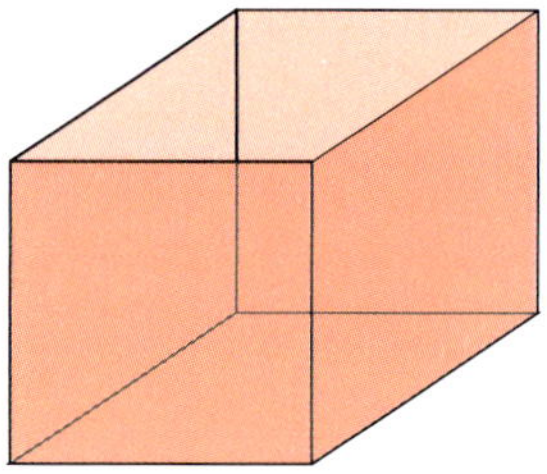 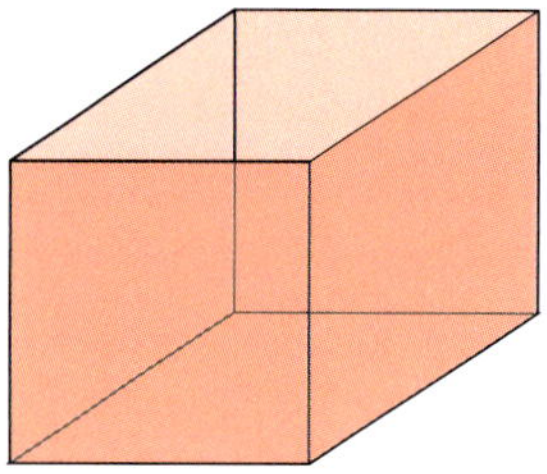 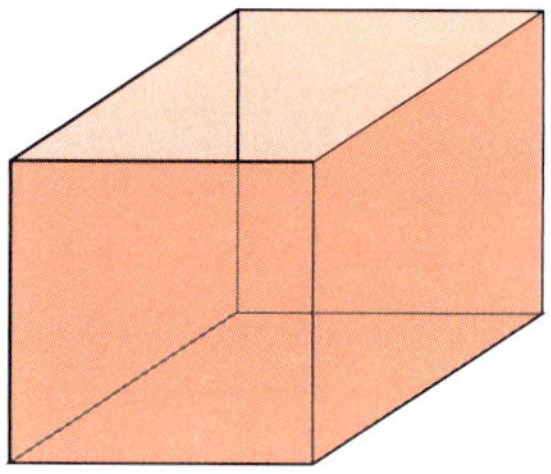

4. How can a piece be broken off and leave behind a **hexagonal cross section**? Use six lines to draw this hexagon.

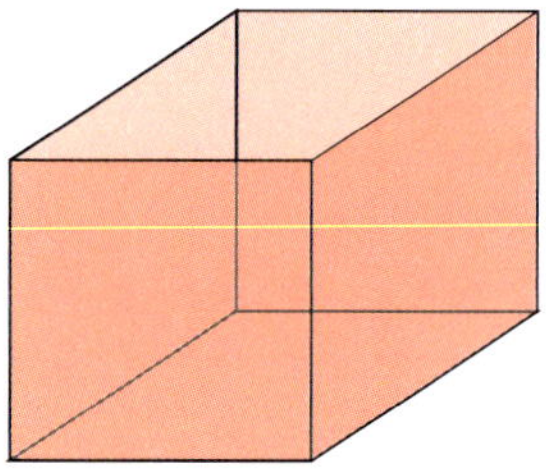 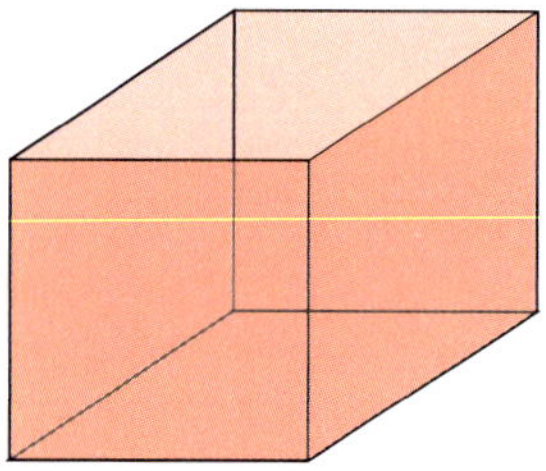 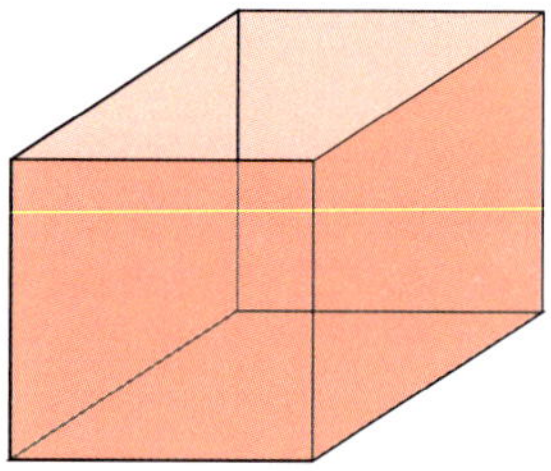

Unit 4: *Cumulative Activity*

In the Hawaiian culture, food was often cooked in underground ovens called imu. The building and use of an imu requires an enormous amount of preparation.

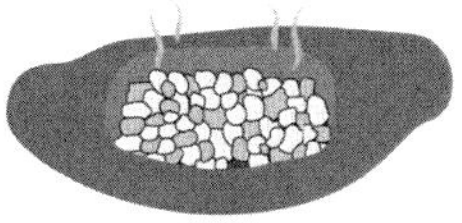

Rectangular imu

Round imu

To build an imu, a pit must be dug into the ground. The pit can have a variety of shapes. Here are some drawings and measurements of common pits used for imu.

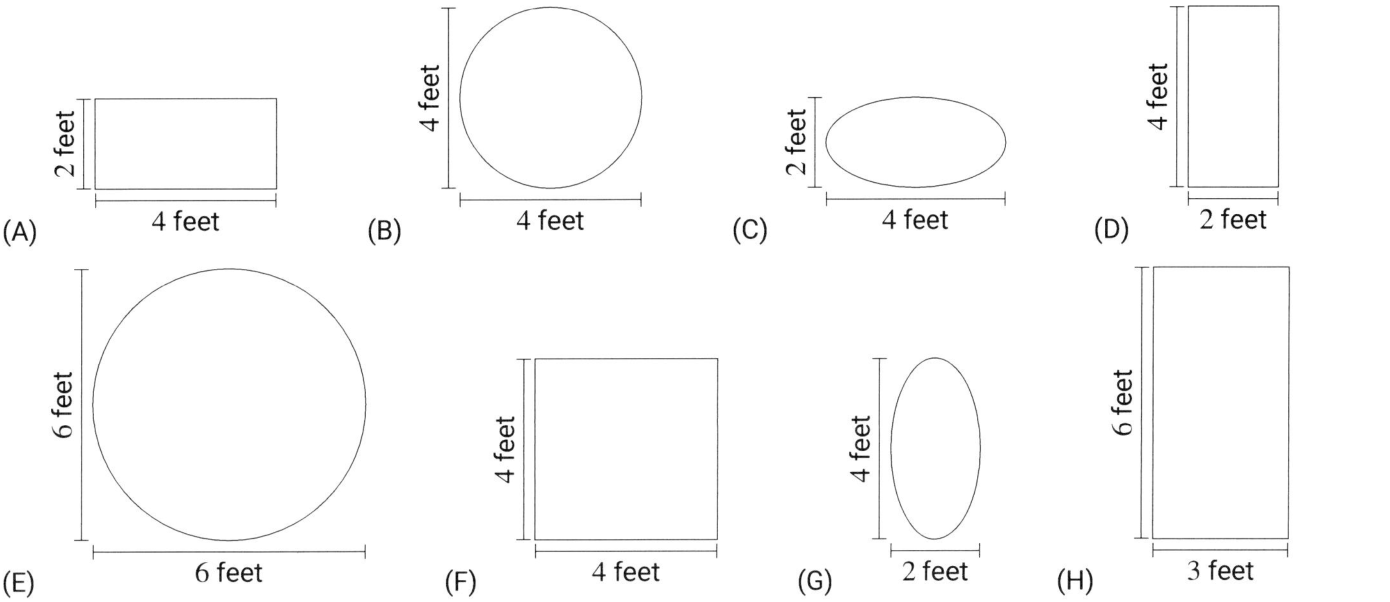

1. Find the perimeter/circumference and area of the two biggest imu in the drawings, (E) and (H). Round to 1 decimal point.

	Perimeter or circumference	Area
Imu (E)		
Imu (H)		

Imu (E)
Circumference $= 2\pi \times 3 \approx 18.9$
Area $= \pi \times (3)^2 \approx 28.3$

Imu (H)
Perimeter $= 3 + 3 + 6 + 6 = 18$
Area $= 3 \times 6 = 18$

2. Mathematically speaking, which of the previous imu drawings are **similar** to this one:

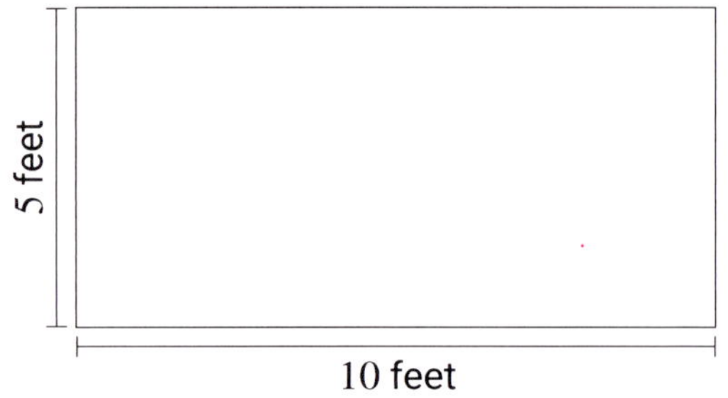

The other shapes that are mathematically similar are the rectangles with one side twice as long as the other. These are Imu (A), (D), and (H).

3. How do you know that the shapes that you chose were similar shapes?

After digging the pit, we build a fire to heat up some rocks. The red hot rocks are then arranged to make space available in the hole.

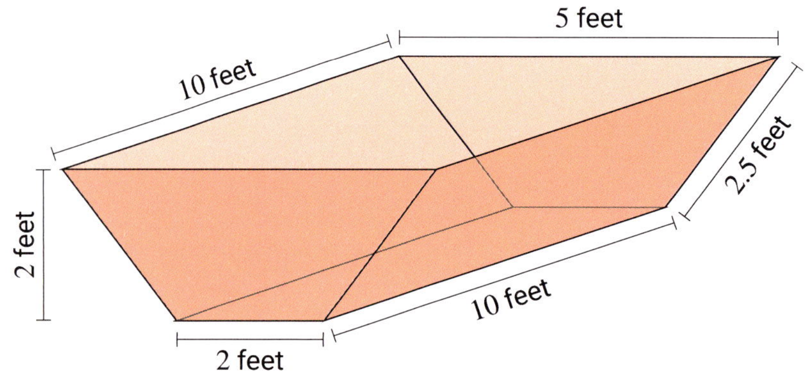

Let's look inside the imu. Our imu is 2 feet deep. The opening of the imu is 5 feet wide by 10 feet long, and the bottom of the imu is 2 feet wide and 10 feet long. The shape turns out to be a trapezoidal prism with the sides sloping in 2.5 feet from the top to the bottom.

4. What is the volume of our imu?

5. Do you think an adult pig would fit in this imu? Discuss with your partner and justify your reasoning. What did you decide and why?

6. Sketch this imu as a geometric **net** (include the rectangle at the opening on top), and label each side of the net. What is the surface area of our imu?

Banana tree stumps, ti leaves, and other native plants are placed on the hot rocks first. This green leafy layer produces steam to cook the food, while protecting the food from the rocks. Meat and other foods are placed on top of the green leafy layer. If a large animal like a whole pig is to be cooked, hot rocks are also placed in their belly. Another layer of banana leaves, old lauhala mats, and tarps cover the food to protect it from dirt. A final layer of dirt will help to keep the heat in until all the food has completely cooked.

7. What would you put in this imu? Below is a cross section of the imu. Draw and label each layer with the items above and anything else you would like to put in there.

8. With your partner or in the online comment section, share some of the ideas you came up with in part 7.

Unit 5: Statistics

In this unit, we'll learn how to use statistics to describe data and make conclusions through fishing with the aunties, exploring traditional Hawaiian measures, and tagging fish. There are three activities in this unit. *Module 10* involves evaluating a fish population through the use of random samples and populations. Module 11 focuses on analyzing and comparing data to help farmers with their crops. The final activity is cumulative and incorporates concepts from each of the previous activities in this unit.

Some of the activities in this unit will require a coin and a six-sided die to complete.

Common Core State Standards

Common Core State Standard	Module 10	Module 11	Unit 5
7.RP.2c Represent proportional relationships by equations. For example, if total cost t is proportional to the number n of items purchased at a constant price p, the relationship between the total cost and the number of items can be expressed as t = pn.	X		X
7.SP.1 Understand that statistics can be used to gain information about a population by examining a sample of the population; generalizations about a population from a sample are valid only if the sample is representative of that population. Understand that random sampling tends to produce representative samples and support valid inferences.	X		
7.SP.2 Use data from a random sample to draw inferences about a population with an unknown characteristic of interest. Generate multiple samples (or simulated samples) of the same size to gauge the variation in estimates or predictions. For example, estimate the mean word length in a book by randomly sampling words from the book; predict the winner of a school election based on randomly sampled survey data. Gauge how far off the estimate or prediction might be.	X		
7.SP.3 Informally assess the degree of visual overlap of two numerical data distributions with similar variabilities, measuring the difference between the centers by expressing it as a multiple of a measure of variability. For example, the mean height of players on the basketball team is 10 cm greater than the mean height of players on the soccer team, about twice the variability (mean absolute deviation) on either team; on a dot plot, the separation between the two distributions of heights is noticeable.		X	X
7.SP.4 Use measures of center and measures of variability for numerical data from random samples to draw informal comparative inferences about two populations. For example, decide whether the words in a chapter of a seventh-grade science book are generally longer than the words in a chapter of a fourth-grade science book.		X	X

Module 10: Random Samples and Populations *Activity*

For this activity, you will need a coin and a six-sided die.

Aunty Momona and Aunty Manini are studying the fish population on the shores of Nānākuli Beach. They both decide to throw their nets into the water a few times to get a random sample of fish. A random sample will allow them to get a rough idea of the fish sizes in this area.

Halfway through the experiment, they realize something is wrong. Aunty Momona's **smallest fish** is much bigger than **almost all** of Aunty Manini's fish!

Aunty Manini suspects that Aunty Momona's sample is not random, but is a biased sample instead.

Aunty Manini Aunty Momona

1. What do you think is actually going on? (Hint: look at the picture of the two aunties fishing.)

The Aunts finally figured out what was disrupting the samples and decide to only use the data that Aunty Manini collected. Below is the data; it shows the length of every fish Aunty Manini caught (in inches).

columns (rows)

7	8	5	1	4	6
8	4	8	13	16	7
15	7	10	18	12	6
5	8	5	12	12	4
12	3	7	6	5	5
10	9	10	10	10	8

(heads)

columns (rows)

10	6	17	9	11	16
10	4	13	11	9	6
5	7	2	7	12	9
8	3	10	15	10	11
1	8	3	7	5	9
13	6	7	10	1	5

(tails)

2. Let's use the coin and die to help us choose some random numbers from the data.
   (a) First flip a coin. If it is "heads", use the table on the left. If it is "tails", use the table on the right.
   (b) Roll a die. This number will tell you which row (from the top) to look at.
   (c) Roll a die again. This number will tell you which column (from the left) to look at.
   (d) Write down your numbers. For example, if you got "tails, 3, 4" then you write down "7." This is because on the right (tails) table, 3rd row, 4th column, it shows that Aunty Manini caught a 7-inch fish.
   (e) Repeat these steps until you have collected 10 numbers. Record your data (numbers) on the tables on the following page.

Coin flip	First die	Second die	Number

Coin flip	First die	Second die	Number

3. Find the mean of the 10 numbers you collected in part 2.

4. The actual mean of all 72 numbers in the table is $8.222...$. How does the mean that you found compare to the actual mean? Is it greater than or less than the actual mean? Is it close to the real mean or way off? Why? How do you think you can get a closer answer? Feel free to discuss with a partner.

5. Let's go back to the full table of 72 numbers.

   (a) What fraction of the fish in Aunty Manini's random sample are greater than or equal to 15 inches?

   (b) If you swam by $2,400$ random fish at this beach. How many of them would you expect to be greater than or equal to 15 inches?

Module 11: Analyzing and Comparing Data *Activity*

In a lo'i, the spacing between the kalo is very important. There are many issues that can occur if they are grown too far apart or too close together. If the kalo are too far apart, you would be wasting land that could be used to grow more plants. If the kalo are grown too close together, it may be hard for the farmer to come into the lo'i and take care of the plants. It would also be easy for bugs and weeds to grow quickly, destroying the kalo.

Traditionally, Hawaiians used their body parts to measure things. In particular, the rows of a lo'i are often spaced one muku apart.

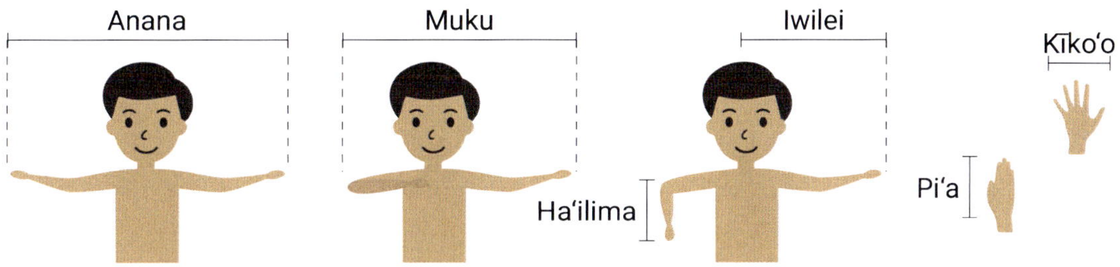

Nā Anakahi Hawai'i: The Hawaiian Units of Measurement

Measuring with our body parts is more convenient than carrying around a large ruler. However, it might cause problems if the farmers have very different body sizes.

Let's take a look at a group of farmers from **Wai'anae** and the lengths of their muku in inches (in).

50 , 56, 46, 42, 52, 38, 38, 46, 42, 50

We will first put the numbers on the dot plot.

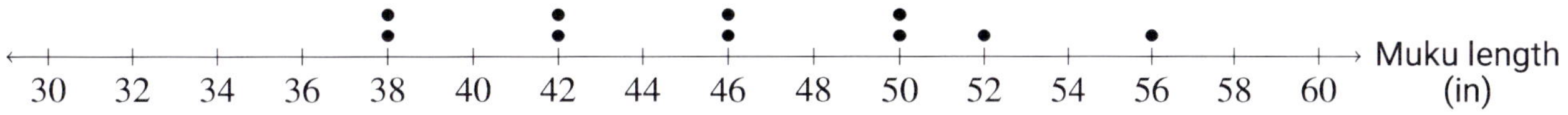

Next, we can find the least value, or minimum, and greatest value, or maximum. Then we find the median, and the lower and upper quartiles to make a box plot.

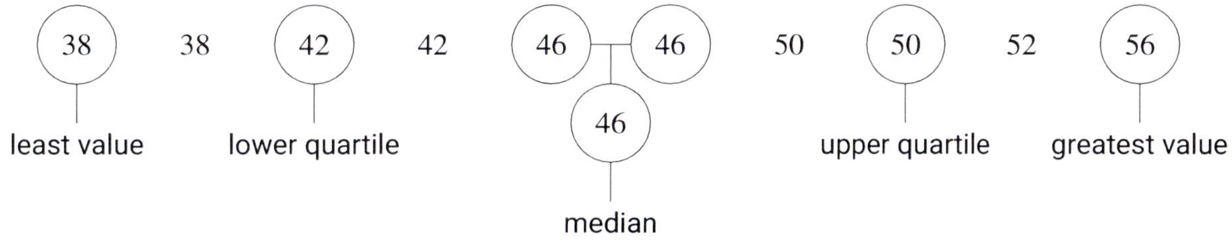

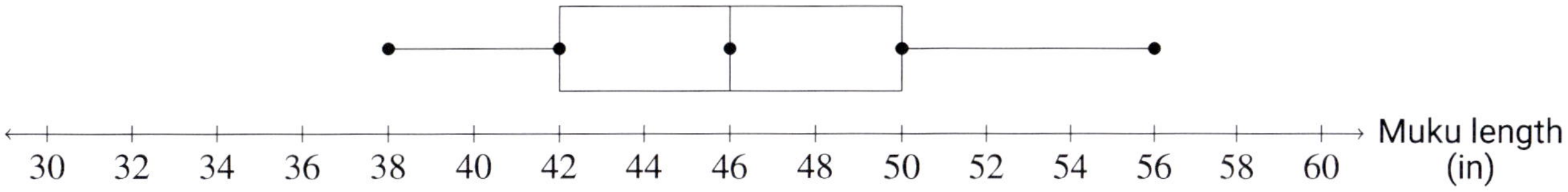

Finally, we can see that the mean is 46:

$50 + 56 + 46 + 42 + 52 + 38 + 38 + 46 + 42 + 50 = 460$

$460 \div 10 = 46$

and the mean absolute deviation (MAD) is 4.8.

$\lvert 50-46\rvert = 4$	$\lvert 56-46\rvert = 10$	$\lvert 46-46\rvert = 0$	$\lvert 42-46\rvert = 4$	$\lvert 52-46\rvert = 6$
$\lvert 38-46\rvert = 8$	$\lvert 38-46\rvert = 8$	$\lvert 46-46\rvert = 0$	$\lvert 42-46\rvert = 4$	$\lvert 50-46\rvert = 4$

$4+10+0+4+6+8+8+0+4+4 = 48$

$48 \div 10 = 4.8$

A group of volunteers from **Kahuku** measured their muku as well, and below is their data.

42 , 50, 38, 38, 40, 32, 36, 38, 36, 40

1. Create a dot plot and box plot of the Kahuku data.

    (a) Dot plot

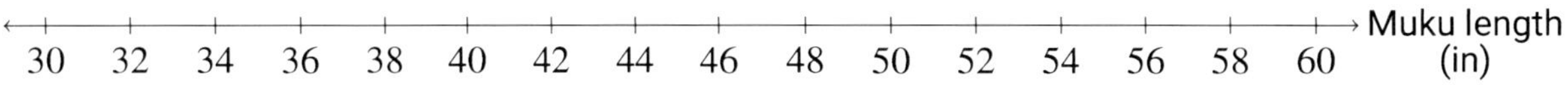

    (b) Box plot

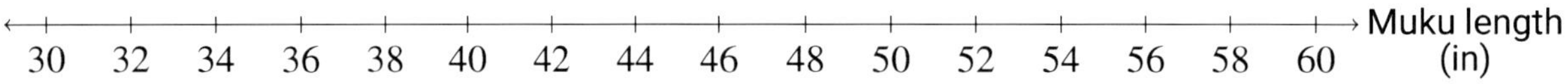

2. Use the plots to compare the Wai'anae farmers with the Kahuku volunteers.

    (a) Compare the shapes of the dot plots.

    (b) Compare the centers of the box plots.

    (c) Compare the spread, or variance, of the plots.

3. Find the mean muku length of the Kahuku volunteers.

4. How many inches longer is the mean muku length of the Wai'anae farmers compared to the mean muku length of the Kahuku volunteers.

5. Using Wai'anae's mean absolute deviation (MAD), how many MADs is Kahuku's mean from Wai'anae's? To find this, divide your answer in part 4 by the MAD of the Wai'anae farmers. Round to one decimal place.

6. Find the MAD of the Kahuku volunteers.

7. Using Kahuku's MAD, how many MADs is Wai'anae's mean is from Kahuku's? To find this, divide your answer in part 4 by the MAD of the Kahuku volunteers in part 6. Round to one decimal place.

Unit 5: *Cumulative Activity*

The ancient Hawaiians were experts in building highly advanced loko iʻa, or fish ponds, which made significant contributions to their sustainable way of life. The loko iʻa were designed to keep fish in, keep their predators out, and allow fresh water to flow through. Unfortunately, most of these fish ponds have disappeared with time, and fish populations are not as they used to be.

In the 2000's, the Hawaiʻi Department of Land and Natural Resources set out to monitor and learn more about the papio fish and the ʻulua (papio over 10 lbs). Fishermen were recruited to help the project by catching, tagging, and releasing these fish. The initial fish length, type, location, and date tagged was recorded by the fishermen and sent to the state. If a tagged fish was caught again, the new length, date, and location was recorded again and sent to the state. With all this data, the state was able to gather a tremendous amount of information about the ulua and papio, giving us an insight into the health of the fish population.

One method of surveying fish populations involves tagging a species of fish, releasing them, and fishing the same area again at a later date. Scientists can then use proportions to make estimates of fish populations based on how many of the caught fish were tagged.

1. Initially 500 papio were tagged in a fish pond. At a later date, fishermen caught and released 100 papio and only 10 of them had tags. Using this information, estimate how many total papio are in the pond. Please explain your reasoning, and show your work.

2. If you know that 25% of mullet's in the fish pond are tagged, and you catch 20 mullet's, how many of those 20 would you expect to have tags? Please explain your reasoning.

We mentioned that fishermen take measurements when they tag fish. Let's compare some 'ulua/papio data that were collected from different fishing spots around O'ahu.

3. Which of these two fishing spots have more large fish? Explain.

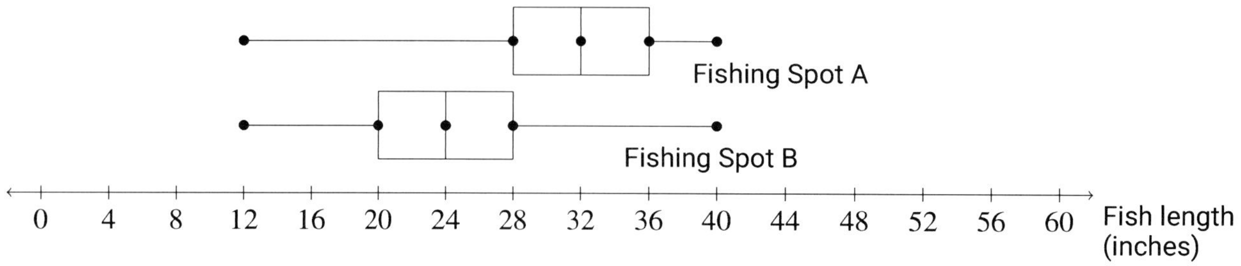

4. Based on the box plots below, which of the two fishing spots would you expect to have more variation in fish sizes? Explain.

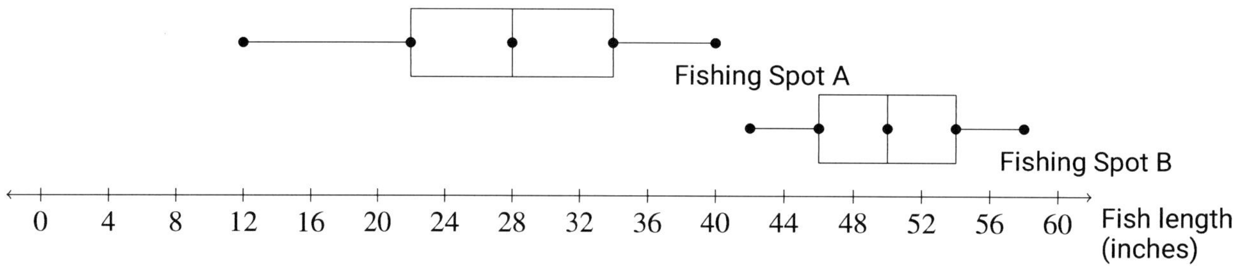

5. Which of these two fishing spots would you rather fish in? Explain what the size of fish you expect to catch and why you chose that fishing spot.

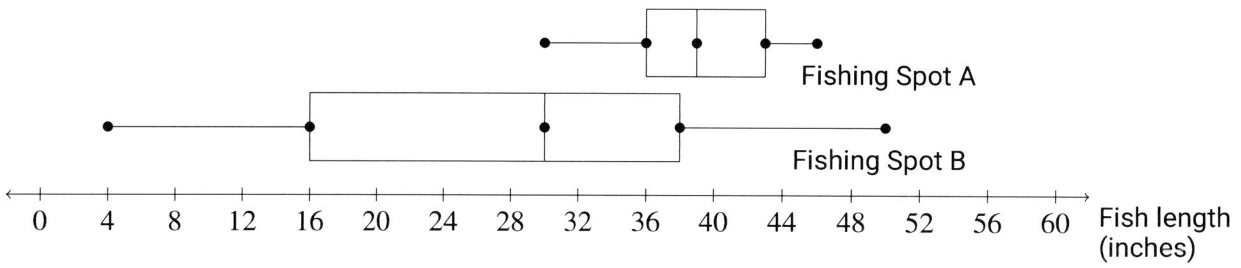

6. With a partner or in the online comment section, explain why you think it is important to do experiments like these. What can we learn from them and how can it help make our world a better place?

Unit 6: Probability

In this unit, we'll learn how to use probability to make predictions and study random events through exploration of the traditional Hawaiian drink, ʻawa, cultivating kalo, and making a plate lunch. There are four activities in this unit. *Module 12* involves experimental probability by tossing a slipper. *Module 13* evaluates true kava from false kava with the use of theoretical probability and simulations. There are two cumulative activities in this unit. Each of the cumulative activities incorporate concepts from each of the previous activities in this unit.

For some of the activities in this unit, students will need a slipper or a flip-flop.

Common Core State Standards

Common Core State Standard	Module 12	Module 13	Unit 6
7.SP.5 Understand that the probability of a chance event is a number between 0 and 1 that expresses the likelihood of the event occurring. Larger numbers indicate greater likelihood. A probability near 0 indicates an unlikely event, a probability around 1/2 indicates an event that is neither unlikely nor likely, and a probability near 1 indicates a likely event.	X	X	X
7.SP.6 Approximate the probability of a chance event by collecting data on the chance process that produces it and observing its long-run relative frequency, and predict the approximate relative frequency given the probability. For example, when rolling a number cube 600 times, predict that a 3 or 6 would be rolled roughly 200 times, but probably not exactly 200 times.	X		X
7.SP.7 Develop a probability model and use it to find probabilities of events. Compare probabilities from a model to observed frequencies; if the agreement is not good, explain possible sources of the discrepancy.		X	X
7.SP.7a Develop a uniform probability model by assigning equal probability to all outcomes, and use the model to determine probabilities of events. For example, if a student is selected at random from a class, find the probability that Jane will be selected and the probability that a girl will be selected.	X	X	X
7.SP.7b Develop a probability model (which may not be uniform) by observing frequencies in data generated from a chance process. For example, find the approximate probability that a spinning penny will land heads up or that a tossed paper cup will land open-end down. Do the outcomes for the spinning penny appear to be equally likely based on the observed frequencies?	X		X
7.SP.8 Find probabilities of compound events using organized lists, tables, tree diagrams, and simulation.	X		X

Continued on next page

Common Core State Standard	Module 12	Module 13	Unit 6
7.SP.8a Understand that, just as with simple events, the probability of a compound event is the fraction of outcomes in the sample space for which the compound event occurs.	X		X
7.SP.8b Represent sample spaces for compound events using methods such as organized lists, tables and tree diagrams. For an event described in everyday language (e.g., "rolling double sixes"), identify the outcomes in the sample space which compose the event.			X
7.SP.8c Design and use a simulation to generate frequencies for compound events. For example, use random digits as a simulation tool to approximate the answer to the question: If 40% of donors have type A blood, what is the probability that it will take at least 4 donors to find one with type A blood?	X		X

Module 12: Experimental Probability *Activity*

In this activity, you will need a slipper or a flip-flop.

Let's take a look at the fishing spots, A, B, C, D, and E. At these spots, there are white fish and there are red fish.

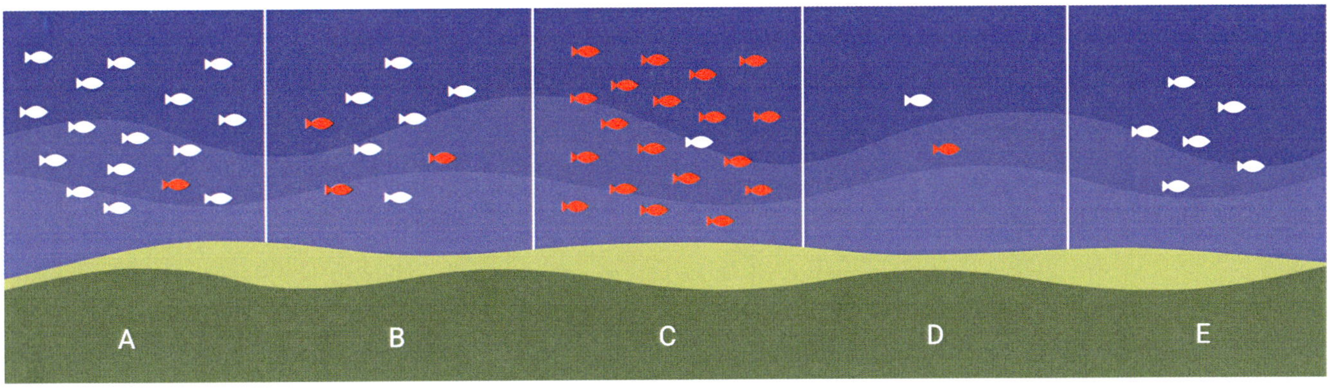

Fishing spots

1. Look at the picture above and match each fishing spot with the probability of a red fish being caught. Use a line to connect each fishing spot with one item on the column to the right that best matches that fishing spot. There will be extra ones on the right that do not match with anything.

Fishing spot	Probability that a caught fish would be red
A•	•Less than 0
	•Exactly 0
B•	•A little more than 0
	•Between 0 and $1/2$
C•	•Close to or equal to $1/2$
	•Between $1/2$ and 1
D•	•A little less than 1
	•Exactly 1
E•	•More than 1

You want to catch red fish, so you go to the spot that gives you the best chance of catching red fish. However, another fisherman arrives at the same time. To be fair, you decide to flip a coin to see who will get to stay at the spot. But... neither one of you have any coins, so you decide to flip something else. What about a slipper?!

Top up

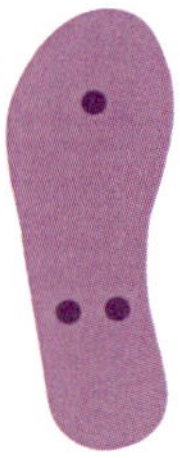

Bottom up

2. Let's get a slipper and toss it gently in the air twice. Be sure to toss the same slipper two times and not a pair of slippers one time. Then, we will note whether it lands with its top up both times, top then bottom, bottom then top, or bottom up both times. Do this 24 times. Use the table below or one like it to create a frequency chart.

		First toss	
		Top up	Bottom up
Second toss	Top up		
	Bottom up		

3. Did each of the possible events occur the same number of times? Why or why not?

4. Use your data to calculate the experimental probabilities of each of the four possible events. Write your results as a simplified fraction.

   (a) Top up both times

   (b) Top up first then bottom up

   (c) Bottom up first then top up

   (d) Bottom up both times

5. If you were to do this experiment a total of 1200 times, based on your results in part 4, about how many times do you expect "top up both times" as a result? Please show your work.

There are several ways to use your slipper to decide who "wins" the fishing spot.

6. If you wanted to make this as **fair** as possible, which of these contests should you choose and why? How do the results of your experiment show that this choice is more fair? (Circle your choice then explain.)
   (a) **Both top up** versus **both bottom up**, otherwise, try again.
   (b) **Top up then bottom up** versus **bottom up then top up**, otherwise, try again.

7. If you wanted to make this as **unfair** as possible, which of these contests should you choose and why? How do the results of your experiment show that this choice is less fair? (Circle your choice then explain.)
   (a) **Both top up** versus **both bottom up**, otherwise, try again.
   (b) **Top up then bottom up** versus **bottom up then top up**, otherwise, try again.

8. Play the **unfair** game from part 7 with a partner. Did you end up having an advantage? Was it a small advantage or a large one?

Module 13: Theoretical Probability and Simulations *Activity*

Kava is an important plant in Polynesia. Originally from Marquesas Islands and Tonga, kava means "bitter" and is also known as ʻawa in Hawaiʻi and ʻava in Samoa. The root can be chewed or ground up and pounded into a saucy texture. After that, the root is mixed with cold water or other ingredients. Kava can be consumed as part of a meal, used as medicine, or as an important part of a ceremony.

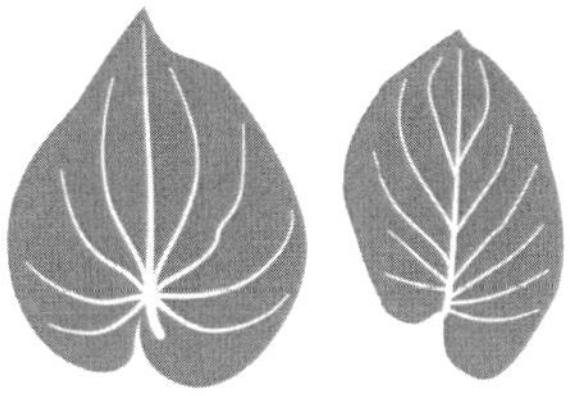

Kava False kava

Unfortunately, there is another plant that looks like kava but that doesn't act like it. This impostor, called the false kava, does not have any of the medical properties of real kava, and it grows much more aggressively. The false kava will spread twice as fast as real kava, covering other plants and taking over entire areas.

Two (2) nature experts and ten (10) student volunteers decide to hike through Nāhiku on Maui to pull out the false kava. However, there are some true kava, and although the experts can easily tell them apart, the volunteers can't.

The experts have found five (5) areas where kava is growing. Here is a rough map of where the true kava (labeled "T") and the false kava (labeled "F") are in each area. Each area had 16 total plants.

Area A Area B Area C Area D Area E

1. For each area, if you were to pull a plant by random, what is the probably that it would be a false kava? Give your answer as a number between 0 and 1. It is okay to give your answer as a simplified fraction.

   (a) Area A

   (b) Area B

   (c) Area C

   (d) Area D

   (e) Area E

2. Suppose a volunteer randomly picks a plant area to pull plants.

   (a) What is the probability of choosing an area (A, B, C, D, or E) that has false kava?

   (b) What is the probability of choosing an area (A, B, C, D, or E) that does not have any false kava?

   (c) What is the probability of choosing Area C and pulling a false kava?

The twelve people will split up and work in different areas at the same time. They want to pull out the false kava while leaving as much of the true kava as possible.

- As we mentioned before, there were two experts who can tell the plants apart and ten volunteers who cannot.
- If an area has at least one expert, they will guide everyone who is with them. All of the false kava will be correctly pulled out, leaving the true kava to thrive.
- If an area only has student volunteers, they will randomly pull out plants since they can't tell the difference between the true and the false kava.

3. How would you distribute the experts and volunteers to each area? Explain your reasoning. The experts and volunteers do not have to be distributed evenly among the areas.

   (a) Area A

   (b) Area B

   (c) Area C

   (d) Area D

   (e) Area E

Unit 6: *Cumulative Activity 1*

Kalo is one of the most important crops the Native Hawaiian people cultivated. In modern times, we make kalo into bread, chips, and even mochi. Traditionally, kalo is pounded into poi, but not all kalo varieties can be used in this process. Only wetland kalo and a few varieties of upland kalo can be made into poi. The difference between wetland and upland kalo is that wetland kalo is grown underwater in a lo'i and upland kalo isn't. There have been over 300 different kalo named in Hawai'i; about half of which is believed to be the same species, just with a different name.

Two important upland kalo are the piko ulaula and the iliuaua. The piko ulaula is great for making poi, and the leaves of the iliuaua are great for laulau.

Piko ulaula Iliuaua

You have to gather a lot of kalo for an upcoming party, so you ask your neighbors for donations. Your neighbors grow a mix of piko ulaula and iliuaua, which they randomly choose to give you.

Neighbor A: 30% of the kalo in this lo'i are piko ulaula.
Neighbor B: 60% of the kalo in this lo'i are iliuaua.
Neighbor C: $1/3$ of the kalo in this lo'i are piko ulaula.
Neighbor D: There are 3 iliuaua for every 1 piko ulaula in this lo'i.

1. Order your neighbors from least likely to most likely to give you a piko ulaula. Show your work and explain.

2. Your neighbors are not giving you the whole kalo plant because they need parts of it to grow more kalo. After randomly choosing a kalo (piko ulaula or iliuaua), they will cut it and randomly give you either the leaves or the main root. This means that you will either get piko ulaula leaves, piko ulaula roots, iliuaua leaves, or iliuaua roots.

   Explain why the model on the right is a more useful model than the one on the left for representing the kalo that make up **Neighbor C**'s lo'i.

Model

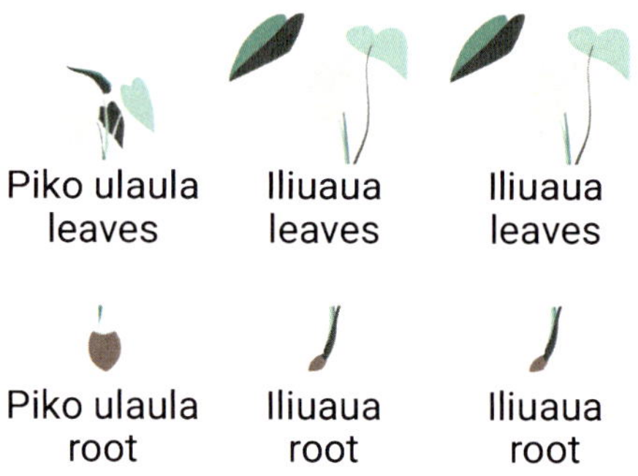

Better model

3. Make a similar model to represent the kalo in **Neighbor D**'s lo'i.

4. The roots of piko ulaula are great for making poi. What is the probability that **Neighbor D** will give you the root of a piko ulaula?

5. The leaves of iliuaua are great for laulau. What is the probability that **Neighbor D** will give you the leaves of a iliuaua?

6. If you receive 64 kalo parts from **Neighbor D**, how many of them do you expect to be the leaves of piko ulaula or the roots of iliuaua? Show your work.

7. With a partner or in the online comment section, make a hypothesis about why one type of kalo is better for poi and the other type is better for laulau.

Unit 6: *Cumulative Activity 2*

In this activity, you will need a six-sided die and a coin.

We are at a lū'au and making ourselves a plate lunch. Let's grab a protein, a plant starch, and a drink.

Proteins	Starches	Drinks
I'a (fish)	Poi (taro)	Coconut water
Moa (chicken)	'Uala (sweet potato)	Māmaki tea
Pua'a (pork)	'Ulu (breadfruit)	

Plate lunch

Let's look at the different kinds of plate lunches that can be made with this menu.

1. Part 1: On the next page, complete the first part of the table by writing all of the possible menu combinations that are missing.

2. Part 2: Let's do an experiment to see what happens when we have to choose our menu randomly.

   (a) Roll a die two times and flip a coin to see which menu combination you would get.

First roll	Protein
1 or 2	I'a
3 or 4	Moa
5 or 6	Pua'a

Second roll	Starch
1 or 2	Poi
3 or 4	'Uala
5 or 6	'Ulu

Coin flip	Drink
Heads	Coconut water
Tails	Māmaki tea

   For example, if you roll and flipped a 3, 6, and tails, then you get a moa-'ulu-māmaki tea combination.

   (b) On the next page, complete the second part of the table by adding a tally for each meal combination that you got in your experiment.

   (c) Repeat until you've gotten **36** meals.

Part 1: Menu combinations			Part 2: Random choices
Proteins	Starches	Drinks	Frequency
I'a	Poi	Coconut water	
I'a	Poi	Māmaki tea	
I'a		Coconut water	
I'a	'Uala		
I'a	'Ulu	Coconut water	
I'a		Māmaki tea	
Moa	Poi	Coconut water	
Moa			
Moa	'Uala	Coconut water	
Moa	'Uala	Māmaki tea	
		Coconut water	
Moa	'Ulu	Māmaki tea	
Pua'a	Poi	Coconut water	
Pua'a	Poi	Māmaki tea	
Pua'a	'Uala	Coconut water	
Pua'a	'Ulu	Coconut water	
Pua'a	'Ulu	Māmaki tea	

Key for Part 2:

First roll	Protein
1 or 2	I'a
3 or 4	Moa
5 or 6	Pua'a

Second roll	Starch
1 or 2	Poi
3 or 4	'Uala
5 or 6	'Ulu

Coin flip	Drink
Heads	Coconut water
Tails	Māmaki tea

Use the table to answer the questions on the following page.

3. How many menu combinations are possible?

4. If you were randomly given 360 meals, how many of each meal combination would you expect to get?

For the following questions, give your answers as reduced fractions.

5. Let's look at the pua'a-'uala-coconut water combination.
   (a) What is the **theoretical** probability that someone would get this menu combination if they chose it randomly?
   (b) Based on your results, what is the **experimental** probability that someone would get this menu combination?

6. Let's look at the menu combinations that have pua'a as the protein.
   (a) What is the **theoretical** probability that someone would get one of these menu combinations if they chose it randomly?
   (b) Based on your results, what is the **experimental** probability that someone would get one of these menu combinations?

7. Let's look at the menu combinations that have māmaki tea as the drink.

   (a) What is the **theoretical** probability that someone would get one of these menu combinations if they chose it randomly?

   (b) Based on your results, what is the **experimental** probability that someone would get one of these menu combinations?

8. How well does the results from your experiment match with the **theoretical** probabilities? Explain why your results did or did not match.

9. Which of the menu combinations would you prefer? Share with a partner or in the online comment section.

Made in the USA
Las Vegas, NV
27 September 2024

95752698R00062